GRAVITY AND SPATIAL INTERACTION MODELS

KINGSLEY E. HAYNES
Director, Center for Urban and Regional Analysis
Indiana University, Bloomington

A. STEWART FOTHERINGHAM
Department of Geography
Indiana University, Bloomington

SAGE PUBLICATIONS
The Publishers of Professional Social Science
Beverly Hills London New Delhi

Printed in the United States of America

For information address:

SAGE Publications, Inc.
275 South Beverly Drive
Beverly Hills, California 90212

SAGE Publications India Pvt. Ltd.
C-236 Defence Colony
New Delhi 110 024, India

SAGE Publications Ltd
28 Banner Street
London EC1Y 8QE, England

International Standard Book Number 0-8039-2326-0

Library of Congress Catalog Card No. L.C. 84-050799

FIRST PRINTING

CONTENTS

INTRODUCTION TO THE SCIENTIFIC GEOGRAPHY SERIES

Scientific geography is one of the great traditions of contemporary geography. The scientific approach in geography, as elsewhere, involves the precise definition of variables and theoretical relationships that can be shown to be logically consistent. The theories are judged on the clarity of specification of their hypotheses and on their ability to be verified through statistical empirical analysis.

The study of scientific geography provides as much enjoyment and intellectual stimulation as does any subject in the university curriculum. Furthermore, scientific geography is also concerned with the demonstrated usefulness of the topic toward explanation, prediction, and prescription.

Although the empirical tradition in geography is centuries old, scientific geography could not mature until society came to appreciate the potential of the discipline and until computational methodology became commonplace. Today, there is widespread acceptance of computers, and people have become interested in space exploration, satellite technology, and general technological approaches to problems on our planet. With these prerequisites fulfilled, the infrastructure needed for the development of scientific geography is in place.

Scientific geography has demonstrated its capabilities in providing tools for analyzing and understanding geographic processes in both human and physical realms. It has also proven to be of interest to our sister disciplines, and is becoming increasingly recognized for its value to professionals in business and government.

The Scientific Geography Series will present the contributions of scientific geography in a unique manner. Each topic will be explained in a small book, or module. The introductory books are designed to reduce the barriers of learning; successive books at a more advanced level will follow the introductory modules to prepare the reader for contemporary developments in the field. The Scientific Geography Series begins with several important topics in human geography, followed by studies in other branches of scientific geography. The modules are intended to be used as classroom texts and as reference books for researchers and professionals. Wherever possible, the series will emphasize practical utility and include real-world examples.

We are proud of the contributions of geography, and are proud in particular of the heritage of scientific geography. All branches of geography should have the opportunity to learn from one another; in the past, however, access to the contributions and the literature of scientific geography has been very limited. I beleive that those who have contributed significant research to topics in the field are best able to bring its contributions into focus. Thus, I would like to express my appreciation to the authors for their dedication in lending both their time and expertise, knowing that the benefits will by and large accrue not to themselves but to the discipline as a whole.

—*Grant Ian Thrall*
Series Editor

SERIES EDITOR'S INTRODUCTION

One of the major intellectual achievements and, at the same time, perhaps the most useful contribution by spatial analysts to social science literature is the development of gravity and spatial interaction models.

In this book, Professors Kingsley Haynes and Stewart Fotheringham provide a clear and comprehensive introduction to these models. This is an excellent and lucid introduction to the evolution of the gravity and spatial interaction models and their specification. The four basic forms of the model are identified and their uses are carefully articulated; the authors outline practical steps to take on the correct specification of these models.

The authors trace the different applications of the gravity model to market area analysis, including the following: determining the boundaries to market areas; determining the demand for a good or service as well as the market threshold for the business; operationalizing the retail model; and outlining the special class of Lowry models. These models are placed within the historical context of the development of the general spatial interaction literature. Haynes and Fotheringham outline the characteristics that have contributed to making these models among the most widely applied in forecasting and in general studies of migration, communications, transportation, and retailing, among other topics in urban and regional analysis.

In a simple and straightforward manner, Haynes and Fotheringham provide real world examples of how these models can be used. Each conceptual development in this book is supported by clear numerical examples that the authors take the readers through step by step. The authors present six fully developed "real world" uses:

(1) planning a new service;
(2) defining retail shopping boundaries;
(3) forecasting migration and voting patterns;
(4) analyzing university enrollments by state;
(5) determining the optimal size of a shopping development; and
(6) locating a facility for maximum patronage.

In addition to outlining the data requirements of the gravity and spatial interaction models, Haynes and Fotheringham finish with a discussion of where one can find published data that may be used to operationalize this model.

Throughout this work, Haynes and Fotheringham keep the discussion at an elementary mathematical level. While their book is primarily intended for readers who are uninitiated to the mysteries of the gravity and spatial interaction models, those familiar with the models will also find that this book is an important contribution to a conceptual organization of the literature. This book will be of special interest to (but is not limited to) students and researchers in the fields of human geography, business and commerce, sociology, demography, political science, archaeology and anthropology, city and regional planning, transportation engineering, regional economics, and regional science.

—*Grant Ian Thrall*
Series Editor

GRAVITY AND SPATIAL INTERACTION MODELS

KINGSLEY E. HAYNES
Director, Center for Urban and Regional Analysis
School of Public and Environmental Affairs
Indiana University, Bloomington

A. STEWART FOTHERINGHAM
Department of Geography
University of Florida, Gainesville

1. GRAVITY MODEL: OVERVIEW

Spatial interaction is a broad term encompassing any movement over space that results from a human process. It includes journey-to-work, migration, information and commodity flows, student enrollments and conference attendance, the utilization of public and private facilities, and even the transmission of knowledge. Gravity models are the most widely used types of interaction models. They are mathematical formulations that are used to analyze and forecast spatial interaction patterns.

The gravity model as a concept is of fundamental importance to modern scientific geography because it makes explicit and operational the idea of relative as opposed to absolute location. All things on the face of the earth

AUTHORS' NOTE: We would like to thank John M. Hollingsworth, Department of Geography, Indiana University, for his excellent cartographic work and patience in developing the diagrams.

can be located in absolute terms by longitude and latitude coordinates, and the absolute position of things can be related to each other by reference to such coordinates. Distances can be specified in these absolute terms. It is then possible to talk about one location as being "five miles from New York City" and another as being "five miles from Bloomington, Indiana." In absolute terms, these two locations are equal in that they are both five miles from an urban area. In relative terms, however, these locations are significantly different in a multitude of ways (for example in terms of access to shopping, access to job opportunities, access to museums and theaters, access to rural life-styles, or access to wilderness opportunities). Each of these significantly differentiates absolute location from relative location. The gravity model allows us to measure explicitly such relative location concepts by integrating measures of relative distance with measures of relative scale or size.

The importance of the relative location concept and spatial interaction can be seen in the application and refinement of the gravity model over the past fifty years. Its continued use by city planners, transportation analysts, retail location firms, shopping center investors, land developers, and urban social theorists is without precedent. It is one of the earliest models to be applied in the social sciences and continues to be used and extended today. The reasons for these strong and continuing interests are easy to understand and stem from both theoretical and practical considerations.

Social scientists are interested in discovering fundamental and generalizable concepts that are basic to social relationships. One of the distinguishing features of human behavior is the ability to travel or move across the face of the earth and to exchange information and goods over distance. Such exchange processes are referred to generically as interaction, and that which occurs over a distance occurs over space. Hence, the general term "spatial interaction" has been developed to characterize this common type of geographic behavior. Shopping, migrating, commuting, distributing, collecting, vacationing, and communicating usually occur over some distance, and therefore are considered special forms of this common social behavior—spatial interaction. We seek here to describe fundamental characteristics that underlie all these forms of social behavior. We will make generalizations about those characteristics that explain or predict similar geographic behavior. Our goal is to demonstrate that spatial interaction models can be considered as the basis of important and useful social theories. The gravity model is one example of a spatial interaction model.

The gravity model, which derives its name from an analogy to the gravitational interaction between planetary bodies, appears to capture and interrelate at least two basic elements: (1) scale impacts: for example, cities with large populations tend to generate and attract more activities than cities with small populations; and (2) distance impacts: for example, the farther places, people, or activities are apart, the less they interact.

These concepts are used by urban social analysts to explain why land values are high in the central areas of cities and at other easily accessible points (Hansen 1959) and why land values are higher in larger cities than in smaller cities. They are used to explain why some public service or retail locations attract more users or customers than do others and to explain the way in which shopping centers impact the areas about them in terms of traffic and customer flows. On a larger scale, they are used to explain the movement of population in the form of migrants, visitors, business and commercial travelers, and the movement of information in the form of mail, telecommunications, and data transfers. In practical terms, these are important topics for many kinds of decision makers, both public and private; a model that purports to reduce the risk in making large capital decisions related to these topics obviously is valuable.

The applications to which the gravity model has been put are not limited to transportation, marketing, retailing, and urban analysis. In archaeology Hallam, Warren, and Renfrew (1976) have used it to help identify probable prehistoric exchange routes in the western Mediterranean, while Jochin (1976) has used it to examine the location and distribution of settlement among hunting-and-gathering peoples. Tobler and Wineberg (1971) used related methods to develop suggestions about the location of lost cities. More recently Clark (1979) has used a type of gravity model analysis to explain archeological data on the flow of goods. In a related context, Kasakoff and Adams (1977) have used location and anthropological information together with a gravity model formulation to explain marriage patterns and clan ties among Tikopians. Trudgill (1975) made a strong argument for the wider use of this method in linguistics, while Hodder (1980) has made a general plea for the wide use of this technique in trying to understand patterns and potential patterns in the spatial organization of both historic and prehistoric activities.

The Model

Figure 1.1 illustrates the basic relationships inherent in gravity models. Compare the expected level of flows between city x and city y with that between cities x and z. Without further information our intuitive expectation would be that the flows between x and y would be larger; y and z are the same distance from x—800 miles—but y has a population of 2 million while z has a population of only 1 million. If interaction were a function of pairs of individuals in any two cities then the potential sets of pairs between x and y is larger than between x and z (2,000,000 × 2,000,000 vs. 2,000,000 × 1,000,000). The potential pairs of interactions would be twice as great in the case of x and y than in the case of x and z. This is the *multiplicative* impact of scale on interaction.

The impact of distance can be demonstrated by comparing the expected levels of flows between x and y, and x and q. The sizes of y and q are the same, so scale is constant. However, without further information we would expect more flows between x and q than between x and y because we would expect the flows between any two points to decline as distance increases. If this decline is proportional to distance, then with scale held constant we would expect half as much interaction between x and y as between x and q.

To generalize, let the scale of each city, population, be represented by P, and the distance between cities be represented by d. Each pair of cities is designated by the subscripts i and j. Interaction between any pair of cities is specified as T_{ij}. The interaction can be expressed as a ratio of the multiplied populations over the distance between any pair of cities,

$$T_{ij} = P_i P_j / d_{ij} \qquad [1.1]$$

Modifications

Three fundamental modifications need to be made to the basic model in equation 1.1. First, the distance element is adjusted by an exponent to indicate whether the impact of distance is proportional or not. For example, the cost per mile of traveling may decrease with distance, as in air travel. Obviously the operational effect of distance would therefore not be directly proportional to airline miles and the negative aspect of distance would need to be reduced or dampened so that the model properly reflects its effects. On the other hand the effect of distance may be underestimated by mileage because the opportunity to know people in cities far away may be reduced by language, culture, and information. The impact of distance may be greater than that indicated by use of straight line mileage in the model. This "distance decay" or "friction of distance" effect will vary depending on the flows being examined—air transportation as opposed to private automobile transportation, for example. Even though distance will always have a negative influence on interaction, in some cases it may be

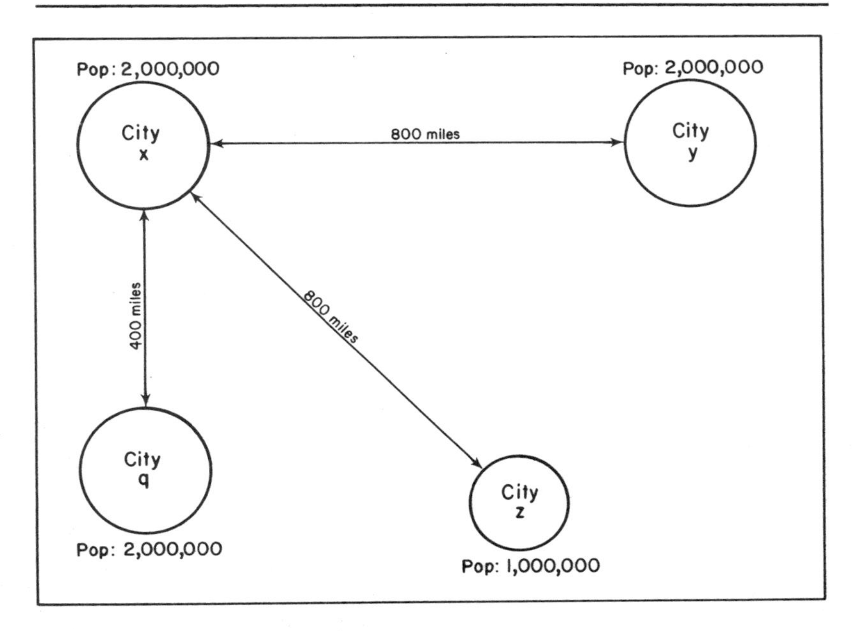

NOTE: This illustration demonstrates the basic principles and trade-offs between the effects of scale (population) and distance (mileage) upon expected interactions between places.

Figure 1.1 The Gravity Model Principles

more negative than in others. An exponent on the distance variable, d_{ij}^{β}, allows us to represent this variability.[1] A large theoretical and empirical literature has developed around the definition of the "correct" exponent.

Much of the literature that focused on deriving the correct exponent for the gravity model formulation was stimulated by physical science interpretations, including the Newtonian analogy where the square of distance, d_{ij}^2, is the appropriate power function. In empirical analysis, however, the exponent is generally interpreted as the responsiveness of interaction to spatial separation and is expected to vary in terms of social context. Larger exponents indicate that the friction of distance becomes increasingly important in reducing the expected level of interaction between centers. Figure 1.2, part a, indicates the impact of small and large values of the exponent β on the distance variable. Other things being equal we would expect distance to have less of an effect in reducing interaction between places i and j in wealthy countries (small β) than in poor countries (large β).

Second, similar arguments can be made for adding an exponent to the population or mass variables, P_i and P_j. The purpose of these exponents is to allow for situations where other variables, aside from population, affect the generation and attraction of interactions. For example, if we were examining the flow of shopping expenditures between two centers we would expect the flow of expenditures to be related not only to population at both centers but also to the average income level at each center. We would expect higher income centers to have greater expenditure flows than do lower income centers of comparable size. An exponent on the population or mass variable would allow for this. In the same way, if the flow being examined were the transfer of electronic data, then perhaps per capita education or computers per capita might represent relevant adjustments. At any rate, such flexibility would appear to be useful both in presenting the general form of the model and in operationalizing the model for special uses. In this way we modify P_i and P_j to arrive at P_i^λ and P_j^α by using the exponents λ and α.[2] Negative exponents would indicate that as population increases, interaction decreases, a characteristic that would be unusual in the real world. Positive exponents would indicate that as population increases, interaction increases. The larger these exponents, the greater the effect of population size on interaction.

The effects of these exponents are graphed in Figure 1.2. In part b of Figure 1.2 it can be seen that the larger the λ exponent, the greater will be the effect of population at place i on interaction from place i. Similarly, from Figure 1.2, part c, it can be seen that the larger the α exponent, the greater will be the effect of population at j on interaction to j. If we were measuring telephone calls in a developing country we might find that there is a population bias in the supply of telephone equipment with disproportionately more equipment being available in larger cities than in smaller ones. Hence, people in larger cities have greater access to telephones. Due to this equipment bias, a population measure taken alone would underestimate this access to equipment and hence underestimate the flow of telephone calls between large places while overestimating the flow of calls between small places. The exponents on population would allow us to correct for this. Small exponents would indicate a small bias in favor of larger cities while large exponents would indicate a large bias in favor of larger cities.

Finally, the third modification to equation 1.1 is the addition of a scale parameter or constant, k, to make the overall equation proportional to the

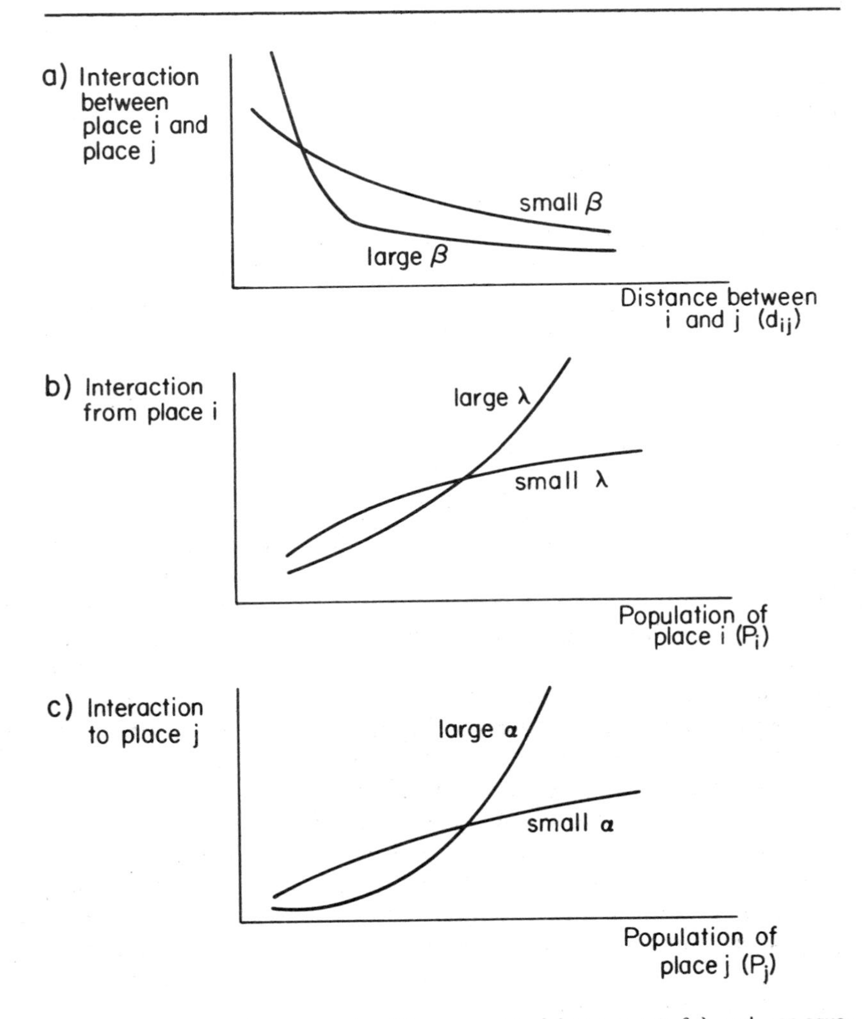

NOTE: These graphs demonstrate the different effects of the exponents β, λ, and α on equation variables distance (d_{ij}), population of origin (P_i) and population of destination (P_j) respectively. Graph a demonstrates the difficulty of overcoming the distance that separates potentially interacting places by indicating that the larger the value of β on d_{ij}^{β} the faster interaction decreases as distance increases. For this reason, β is known as the distance decay parameter. Graph b demonstrates that the larger the exponent λ, the greater the propulsiveness or "sending power" of large places compared to small places. Graph c demonstrates that the larger the exponent α, the larger the attractiveness or "sending power" of large places compared to small places.

Figure 1.2 Population and Distance Exponents

"rate characteristic" of the phenomena we are modeling. Suppose we wish to model airline passenger flows daily and monthly. The population variables, P_i and P_j, and the distance variable d_{ij} will be the same but the magnitude of the flows will be different since monthly flows will be larger than daily flows. Our constant k is used to adjust for such differences in magnitude.

With the three modifications our gravity model equation becomes

$$T_{ij} = k \frac{P_i^{\alpha} P_j^{\alpha}}{d_{ij}^{\beta}} \qquad [1.2]$$

or alternatively,

$$T_{ij} = k\ P_i^{\lambda} P_j^{\alpha}\ d_{ij}^{\beta} \qquad [1.3]$$

The difference between equation 1.2 and equation 1.3 is that in the former we expect β to be positive since we are dividing by d_{ij}^{β} and in the latter we expect β to be negative since we are multiplying by d_{ij}^{β}. It is important to note that equations 1.2 and 1.3 are two identical forms of the same model and can be used interchangeably.

In reality this model has been used more as a basis for explanation than as a basis for prediction but explanation is often a precursor to quality prediction. Furthermore, while this simple model has often proved less reliable than some would desire (due in part to the complex nature of problems it was expected to explain) it has, in recent years, been significantly expanded and refined and its applicability has been substantially widened.

Evolution

A number of alternative labels have been applied to the basic concept in equation 1.2, often reflecting the particular application, estimation procedure, or analytic variation employed in the operation of the model. Variants on the same theme include the following: Reilly's Law of Retail Gravitation (1929); Huff's Model of Consumer Behavior (1959); Dodd's Interactance Hypothesis Model (1950); Zipf's Minimum Effort Model (1949); Stouffer's Intervening Opportunity Model (1940, 1960), and Wilson's Entropy Model (1967). However, the name that continues to be attached most widely to the concept, the gravity model, reflects its origins as an attempt to utilize Newtonian physics to explain social science phenomena.

The Newtonian physics approach to social science analysis continues to influence and cloud the application and interpretation of the model. Unfortunately, it is a line of influence that began more than a century ago with Carey's (1858) attempts to develop social science concepts with physical science principles; it has continued intermittently through the development of empirical applications to migration (Ravenstein 1885; Young 1924) to shopping (Reilly 1929) to undergraduate enrollments (Stewart 1940) and, in recent reformulation and extension, to transportation (Wilson 1967).

The physics analog to the gravity model has plagued attempts to refine and extend the method. Whether it be a deterministic Newtonian approach, a statistical mechanics concept, or an application of the second law of thermodynamics (entropy), the physics analog has muddled the interpretation of empirical social science results and has isolated the work from parallel and alternative theory and analytic procedures in the main stream of social science. Even the sociological approach of Dodd's interactance hypothesis (1950), which replaced the deterministic perspective of Newtonian physics with a more probabilistic base, did not escape interpretation as a particle physics analog. Zipf's principle of least effort, which provided a different and intriguing framework for explaining the gravity model concepts, substituted an equally myopic biosocial analog for the flawed social physics paradigm. It was only with the development of Huff's consumer behavior concepts and Stouffer's intervening opportunity ideas that a theory relating to human behavior was developed for gravity model concepts and interpretations. However, these new conceptual advances ran into the problem of applying individual level explanations of behavior to a model that only describes aggregate outcomes. Even the developments of Niedercorn and Bechdolt (1969, 1972), who derived the gravity model from economic principles of utility maximization, had to address this thorny aggregation issue.

Wilson turned the problem on its head and explicitly introduced statistical mechanics concepts avoiding the problem of aggregation by starting at the macro rather than the micro level in developing estimation procedures for interaction. The use of statistical mechanics and entropy concepts was of concern to many social scientists because it seemed to be yet another social physics analog. This was unfortunate in that Wilson's approach allowed aggregate interaction to be treated as a basic estimation problem in information theory utilizing the entropy concept not as a law in thermodynamics but as a statement of uncertainty in a probability distribution (Jaynes 1957; Webber 1977, 1980). Thus, gravity model concepts were related to basic statistical estimation procedures and Bayesian inferential concepts (March and Batty 1975) as well as to the extremal theory basis of optimization methods (Charnes, Raike, and Bettinger 1972; Haynes and Phillips 1982). As an estimation procedure, this completely removed the model from its social physics framework and interpretively linked it to modern day mathematical statistics (Haynes, Phillips, and Mohrfeld 1980).

Expansion and Generalization

To appreciate this evolution fully, it is vital to examine the basic concepts of the gravity model and see how they have expanded from the original formulation to the highly generalized and widely applicable model that exists today. Expansion occurred through (1) the development of the potential concept; (2) the separation of origin and destination attributes; (3) the identification of intervening opportunity and agglomeration impacts; and (4) a realization of the effects of spatial structure on interactions.

The first expansion simply generalizes the model from a flow between a single pair of centers as indicated in the equation 1.2 to a set of flows among all centers in the system. This total flow estimate in a system of centers is derived by summing the gravity model estimates of all possible pairs (see Figure 1.1) so that, for example,

$$T_{total} = k\frac{P_x^{\lambda}P_y^{\alpha}}{d_{xy}^{\beta}} + k\frac{P_x^{\lambda}P_z^{\alpha}}{d_{xz}^{\beta}} + k\frac{P_x^{\lambda}P_q^{\alpha}}{d_{xq}^{\beta}} + k\frac{P_z^{\lambda}P_y^{\alpha}}{d_{zy}^{\beta}} + k\frac{P_y^{\lambda}P_q^{\alpha}}{d_{yq}^{\beta}} + k\frac{P_z^{\lambda}P_q^{\alpha}}{d_{zq}^{\beta}} \quad [1.4]$$

More commonly, we estimate a set of flows between one center and all other centers of interest in a pair-by-pair process. This single centered subtotal of flows is referred to as a center's interaction potential. In the case of Figure 1.1 assuming $k = 0.000001$, $\lambda = 1.0$, $\alpha = 1.0$ and $\beta = 1.0$, the potential for center x would be as follows:

$$\text{Potential } T_x = k\frac{P_xP_y}{d_{xy}} + k\frac{P_xP_z}{d_{xz}} + k\frac{P_xP_q}{d_{xq}} = 5{,}000 + 2{,}500 + 10{,}000 = 17{,}500$$

The second expansion acknowledges the importance of developing both a careful distinction between origins and destinations and appropriate measures to represent the importance of these variables. P_i and P_j have been used so far to indicate the propulsive (sending) power of i and the attractive (drawing) power of j, respectively. Historically, population has been used for this purpose because it is closely related to many flows we wish to measure. Population, however, is only a surrogate variable; it stands for the propulsive and attractive forces we wish to measure. In analyzing shopping behavior, for example, the drawing power or attraction of a shopping center destination is not its population but rather its size in number of stores or square footage of retail space. Our use of exponents on population was a way of trying to adjust this measure to make it appropriately represent these propulsive (sending) and attractive (drawing) forces. When we can measure these forces directly such as by counts of customers visiting a store in a day or by traffic counts of the number of trips leaving a given suburb in a day we do not need "stand in" or surrogate variables. We simply put in the actual flow from each origin, O_i, and/or to each destination, D_j.

In those cases where O_i and D_j are unknown and where we need to estimate these forces, population is only one specific estimate. If we wish to be general we could say that this estimate of the generating capacity of flows from an origin could be made up of a number of attributes. In the simplest and most general terms we could describe this list or vector of origin-generating flow attributes as $\underline{V}_i$ and the list or vector of destination-attracting flow attributes of as $\underline{W}_j$. In this way, we still distinguish between origins, i, and destinations, j, but we do not imply a specific estimate or individual variable such as population for the propulsive or attractive force. Instead, we might have several surrogate variables for origin propulsiveness and destination attractiveness. For example, in terms of migration within the United States, destination attractiveness could be measured by population, climate, economic variables, and social variables.

The third expansion of the gravity concepts deals with the separation of origins from destinations and the impact of that intervening distance or space. Not only does distance inversely affect the flow of interactions between an origin and a destination, so does the presence of alternative destinations. At the moment, our model would predict the interaction between Boston and Philadelphia to be the same whether New York City were present or not. It would seem reasonable that the intervening nature of New York City would negatively influence the interaction between Boston and Philadelphia. These intervening opportunities reduce the flow from the place of origin to the place of destination by absorbing or diverting interactions to closer destinations. Such an inverse impact represents the negative effect of intervening opportunities on interaction.

Alternatively, opportunities (destinations) that cluster together may have an agglomeration effect drawing more flows from an origin than would be drawn if these destinations were assessed independently. This agglomeration effect is most clearly seen in shopping centers that draw more patrons than would be the case if the stores were dispersed. The reason for this agglomeration effect is that shopping centers are efficient: One stop allows a person to access a whole range of retail opportunities, and reduces travel costs. Therefore, it is useful for our general model that we not rely on one simple measure of the effect of spatial separation of origins and destinations such as distance. Clearly it would be preferred to have a list or vector of variables whose attributes represent the negative impact of spatial separation on interaction (such as distance and intervening opportunities). In order to do this we substitute for distance a general vector, $\underline{S}_{ij}$, that represents the different impacts of spatial separation on interaction between i and j.

A fourth expansion to the gravity model concerns the spatial pattern of origins and destinations. This pattern is generally referred to as spatial structure and can have an impact on spatial interaction patterns, as described in Chapter 4. For the moment, however, our general expanded model thus estimates interaction, T_{ij}, with a set of three vectors,

$$T_{ij} = f(\underline{V}_i, \underline{W}_j, \underline{S}_{ij}) \quad [1.5]$$

where $\underline{V}_i$ represents a vector of origin attributes; $\underline{W}_i$ represents a vector of destination attributes; and $\underline{S}_{ij}$ represents a vector of separation attributes.

NOTES

1. The exponent is sometimes introduced into the model by the function exponential (βd_{ij}). For the sake of brevity, however, this monograph discusses only the function $d_{ij}{}^{\beta}$. It should be noted that the general statements made herein apply equally to both functions.

2. In a similar manner to the distance exponent, λ and α could be introduced in functions such as exponential (λP_i) and exponential (αP_j). These functions are less commonly used than the equivalent distance function and are not discussed further.

2. A FAMILY OF GRAVITY MODELS

The purpose of this chapter is to describe different gravity models that can be generated from the general formula discussed in the previous chapter:

$$T_{ij} = f(\underline{V}_i, \underline{W}_j, \underline{S}_{ij}) \quad [2.1]$$

In what follows we will assume a single measure of origin propulsiveness, destination attractiveness, and spatial separation; namely, each of the vectors $\underline{V}_i$, $\underline{W}_j$, and $\underline{S}_{ij}$ contains only one variable. We will use lower case letters to represent this single measure situation (v_i, w_j, d_{ij}). Remember, however, that there is no restriction on the number of variables that can be used to describe a specific attribute; also, including more than one variable for any or all attributes would not alter the general discussion. For convenience, we assume that spatial separation can be measured accurately by distance and that origin propulsiveness and destination attractiveness can be measured accurately by some size variable such as population. We may, in some instances, have good estimates or even exact values for outflow totals from the origins or inflow totals into destinations. Since we define origin propulsiveness and destination attractiveness in terms of such totals, we would want to include these totals in the gravity model instead of the less accurate size variables. The various ways the size variables are replaced by outflow and/or inflow totals produces a "family" of gravity models (Wilson 1971).

Model Specification

Various types of gravity models can be obtained from the general formula in equation 2.1; the type of model to be used in any particular situation depends on the information available on the interaction system.

INFORMATION AVAILABLE ONLY ON THE TOTAL NUMBER OF INTERACTIONS IN THE SYSTEM

Suppose that we have an accurate estimate of the total number of interactions in a system. We have no other information apart from this and we are asked to forecast the interaction pattern in the system. If we have m origins and n destinations, then obviously we require

$$\sum_{i=1}^{m} \sum_{j=1}^{n} \hat{T}_{ij} = \sum_{i=1}^{m} \sum_{j=1}^{n} T_{ij} = T \qquad [2.2]$$

where $\hat{T}_{ij}$ is the estimated interaction between i and j and T_{ij} is the actual interaction. T is defined as the total number of interactions in the system (the only knowledge we have on such flows). Equation 2.2 represents a constraint that states that we want the sum of the predicted interactions from our gravity model to equal the sum of the actual interactions. A gravity

model containing only this simple constraint is termed a *total flow constrained gravity model*. Its form is

$$\hat{T}_{ij} = k\ v_i^{\lambda} w_j^{\alpha} d_{ij}^{\beta} \qquad [2.3]$$

which is the same as that of equation 1.1 but here v_i represents the origin propulsiveness variable; w_j represents the destination attractiveness variable, d_{ij} represents distance; λ, α, and β represent exponents or parameters to be estimated; and k, the scale parameter, is defined as

$$k = T / \sum_i \sum_j v_i^{\lambda} w_j^{\alpha}\ d_{ij}^{\beta} \qquad [2.4]$$

This formulation ensures that the constraint in equation 2.2 is met.

Although obtaining values for the parameters λ, α, and β is beyond the scope of this book we have a priori expectations about their signs. We expect λ and α to be positive, indicating that as an origin and a destination increase in size, the volume of interaction between them increases. Conversely, we expect β to be negative: As the separation between an origin and a destination increases, the volume of interaction between them decreases. In practice, values of λ and α are often found to be between 0.5 and 2.0, and values of β are often found to be between –0.5 and –2.0.

INFORMATION AVAILABLE ON THE OUTFLOW TOTALS FOR EACH ORIGIN

If the outflow totals are known or can be predicted accurately for each origin in the system, then this automatically implies that the total number of interactions in the system is known or can be predicted. Thus, we have all of the information we had in the preceding situation but now we also have information on the total number of interactions leaving each origin. In Chapter 1 we defined this variable as O_i so that

$$O_i = \sum_j T_{ij} \qquad [2.5]$$

and we require

$$\sum_j \hat{T}_{ij} = O_i \text{ for all } i \qquad [2.6]$$

That is, the predicted total interaction volume leaving each origin should equal the known value, O_i. These "known" values may have been estimated from a trip generation equation (for example, see Hutchinson 1974) or they may be derived from a survey. In both cases we know how many people left a particular origin but we do not know where they went. A task of the gravity modeler is to estimate "where they went" and the appropriate gravity model for this is the *production-constrained gravity model.* The form of this model is

$$\hat{T}_{ij} = A_i O_i w_j^{\alpha} d_{ij}^{\beta} \qquad [2.7]$$

where

$$A_i = [\sum_j w_j^{\alpha} d_{ij}^{\beta}]^{-1} \qquad [2.8]$$

so that the model can also be represented as a "share" model

$$\hat{T}_{ij} = \frac{O_i w_j^{\alpha} d_{ij}^{\beta}}{\sum_j w_j^{\alpha} d_{ij}^{\beta}} \qquad [2.9]$$

In equation 2.7, A_i is termed a balancing factor because it imposes the constraint given in equation 2.6. A_i also measures the relative location of origin i to the destinations j: High values of A_i are associated with origins that are inaccessible; low values are associated with origins that are accessible.

The production-constrained gravity model is useful in forecasting destination inflow totals that are unknown. If $\hat{D}_j$ is the predicted total inflow into j,

$$\hat{D}_j = \sum_i \hat{T}_{ij} \qquad [2.10]$$

A situation where such prediction is useful is in modelng shopping expenditures. The population and average income levels of residential zones are usually attainable, and so an estimate of O_i (the available disposable income for shopping in each zone) can be made. The production-constrained gravity model can then be used to forecast the revenues generated by particular shopping locations. Examples 5 and 6 in Chapter 5 describe this type of application of the model.

INFORMATION AVAILABLE ON THE INFLOW TOTALS FOR EACH DESTINATION

In this case we know the inflow totals into each destination but not the outflow totals for each origin. This then is simply the reverse of the above situation. If inflow totals are known or can be estimated accurately, then we require a gravity model that constrains

$$\sum_j \hat{T}_{ij} = D_j \text{ for all j} \qquad [2.11]$$

where D_j is the known inflow into destination j. Such a model is termed an *attraction-constrained gravity model* and has the following form:

$$\hat{T}_{ij} = v_i^{\lambda} B_j D_j d_{ij}^{\beta} \qquad [2.12]$$

where

$$B_j = [\sum_i v_i^{\lambda} d_{ij}^{\beta}]^{-1} \qquad [2.13]$$

The model can thus be represented as

$$\hat{T}_{ij} = \frac{D_j v_i^{\alpha} d_{ij}^{\beta}}{\sum_i v_i^{\alpha} d_{ij}^{\beta}} \qquad [2.14]$$

In equation 2.12, B_j is known as a balancing factor that ensures that the constraint given in equation 2.11 is met. B_j also measures the relative location of destination j to the origins i: High values B_j are associated with destinations that are inaccessible; low values are associated with destinations that are accessible.

The attraction-constrained gravity model can be used to forecast total outflows from origins. Such a situation might arise, for example, in forecasting the effects of locating a new industrial park within a city. The number of people to be employed in the new development is known, and the attraction-constrained gravity model can be used to forecast the demand for housing in particular parts of the city that will result from the new employment.

One urban geography model that utilizes the properties of both the production-constrained gravity model and the attraction-constrained gravity model is the Lowry model (Lowry 1964). The attraction-constrained model is used to allocate workers to residential zones based on a given distribution of basic industry; the production-constrained model is then used to allocate retail expenditures and retail employment to various zones given the residential pattern. A further description of this model is given in Chapter 3, which discusses retail gravity models and marketing.

INFORMATION AVAILABLE ON BOTH OUTFLOW TOTALS AND INFLOW TOTALS

Suppose we are given the task of forecasting traffic patterns or migration patterns and we know or can estimate accurately the outflow totals for each origin and the inflow totals for each destination. The gravity model chosen for such a task should ensure that both the constraints in equations 2.6 and 2.11 operate. Such a model is termed a production-attraction-constrained gravity model, or alternatively, a *doubly constrained gravity model.* It has the following formulation:

$$\hat{T}_{ij} = A_i O_i B_j D_j d_{ij}{}^{\beta} \qquad [2.15]$$

where

$$A_i = [\sum_j B_j D_j d_{ij}{}^{\beta}]^{-1} \qquad [2.16]$$

and

$$B_j = [\sum_i A_i O_i d_{ij}{}^{\beta}]^{-1} \qquad [2.17]$$

A_i is a balancing factor that ensures the constraint in equation 2.6 and B_j is a balancing factor that ensures the constraint in equation 2.11

In practice, A_i and B_j are estimated iteratively as they are functions of each other. Usually the iteration is started by setting all of the B_j's to 1.0 and deriving initial estimates of the A_i's from equation 2.16. These estimates of the A_i's are then used to obtain estimates of the B_j's via equation 2.17 and the process continues until the values of all the A_i's and B_j's exhibit no change on successive iterations. Masser (1972) gives a worked example of the calculation of the balancing factors in a doubly constrained gravity model.

A Comparison of the Predictive Behavior of the Four Gravity Models

Consider the following 3 × 3 interaction matrix:

	from/to	Destinations (j) 1	2	3	Total Outflows ($\sum_j T_{ij}$ or O_i)
	1	100	20	40	160
Origins (i)	2	60	300	90	450
	3	40	50	90	180
Total Inflows ($\sum_i T_{ij}$ or D_j)		200	370	220	Grand Total 790 ($\sum_i \sum_j T_{ij}$ or T)

where the distance matrix between the three locations is

	1	2	3
1	2	15	5
2	15	2	10
3	5	10	2

Suppose $\lambda=1$, $\alpha=1$, and $\beta=-1$ for this system.

If v_i is measured by O_i and w_j is measured by D_j, then the predicted flow matrix from a *total-flow-constrained gravity model* is

		Destinations (j)			$\sum_j \hat{T}_{ij}$
	from/to	1	2	3	
	1	79	19	35	133
Origins (i)	2	29	412	49	490
	3	36	33	98	167
$\sum_i \hat{T}_{ij}$		144	464	182	790 $\sum_i \sum_j \hat{T}_{ij}$

The value of k in the model that ensures that $\sum_i \sum_j \hat{T}_{ij} = \sum_i \sum_j T_{ij} = 790$ is 0.004944. To demonstrate how the predicted interactions in the above table are derived, consider the predicted flow $\hat{T}_{11}$ which is 79. From equation 2.3, this value is calculated from the expression $kv_1^{\lambda}w_1^{\alpha}d_{11}^{\beta}$, which in this case is equal to $kO_1D_1d_{11}^{-1}$, which is

$$\frac{0.004944 \times 160 \times 200}{2} = 79$$

The reader should attempt to derive the other predicted interactions in a similar manner.

The value of k is derived by calculating $O_iD_jd_{ij}^{\beta}$ for each interaction, summing these values, and dividing the sum into the actual total number of flows in the system, which is 790. Notice in the model predictions that while the predicted grand total is equal to the actual grand total, there is no constraint on either the predicted total outflows ($\sum_j \hat{T}_{ij}$) or on the predicted total inflows ($\sum_i \hat{T}_{ij}$).

The predicted flow matrix from a *production-constrained gravity model* is

	from/to	Destinations (j) 1	2	3	$\sum_j \hat{T}_{ij}$
	1	95	23	42	160
Origins (i)	2	27	378	45	450
	3	38	36	106	180
$\sum_i \hat{T}_{ij}$		160	437	193	790

To see how each of these values is derived, consider the predicted flow $\hat{T}_{11}$, which is 95, and from equation 2.7 is equal to $A_1 O_1 w_1{}^{\alpha} d_{11}{}^{\beta}$, which in this case is equal to $A_1 O_1 D_1 d_{11}{}^{-1}$. The only unknown in the formula is A_1, which from equation 2.8 is equal to

$$A_1 = [\sum_j w_j{}^{\alpha} d_{1j}{}^{\beta}]^{-1} = [\sum_j D_j d_{1j}{}^{-1}]^{-1} \qquad [2.18]$$

so that
$$A_1 = \left[\frac{200}{2} + \frac{370}{15} + \frac{220}{5}\right]^{-1} = [168.67]^{-1} = 0.005929$$

By an identical derivation, $A_2 = 0.004539$ and $A_3 = 0.005348$. Then,

$$\hat{T}_{11} = \frac{0.005929 \times 160 \times 200}{2} = 95$$

and the other $\hat{T}_{ij}$'s can be calculated in the same way.

In the table of predicted flows from the production-constrained gravity model notice that the constraint in equation 2.6 operates and $\sum_j \hat{T}_{ij} = \sum_j T_{ij} = O_i$ for each origin, but there is no constraint on inflow totals.

The predicted flow matrix from an *attraction-constrained gravity model* is

	from/to	Destinations (j) 1	2	3	$\sum_j \hat{T}_{ij}$
	1	110	16	42	168
Origins (i)	2	41	328	59	428
	3	49	26	119	194
$\sum_i \hat{T}_{ij}$		200	370	220	790

To see how each of these values is derived, consider the predicted flow $\hat{T}_{11}$, which is 110, and from equation 2.12 is equal to $v_1{}^{\lambda} B_1 D_1 d_{11}{}^{\beta}$, which in this case is equal to $O_1 B_1 D_1 d_{11}{}^{-1}$. B_1 is calculated from equation 2.13 and is equal to

$$B_1 = [\sum_i v_i{}^{\lambda} d_{i1}{}^{\beta}]^{-1} = [\sum_i O_i d_{i1}{}^{-1}]^{-1} \quad [2.19]$$

so that

$$B_1 = \left[\frac{160}{2} + \frac{450}{15} + \frac{180}{5}\right]^{-1} = [146]^{-1} = 0.006849$$

By an identical derivation, $B_2 = 0.003942$ and $B_3 = 0.005988$. Then,

$$\hat{T}_{11} = \frac{0.006849 \times 160 \times 200}{2} = 110$$

and the other T_{ij}'s can be calculated in the same way.

In the table of predicted flows from the production-constrained gravity model, notice that the constraint in equation 2.11 operates and $\sum_i \hat{T}_{ij} = \sum_i T_{ij} = D_j$ for each destination but there is no constraint on the outflow totals.

Finally, the predicted flow matrix from a doubly constrained gravity model is as follows:

		Destinations (j)			$\sum_j \hat{T}_{ij}$
	from/to	1	2	3	
	1	107	13	40	160
Origins (i)	2	47	334	69	450
	3	46	23	111	180
$\sum_i \hat{T}_{ij}$		200	370	220	790

To derive each of these values it is first necessary to compute the A_i's and B_j's as they are defined in equations 2.16 and 2.17, respectively. This is best done on a computer, and the calculation is not shown by hand. The values for A_1 and B_1 are 0.0046 and 1.45 respectively, so $\hat{T}_{11}$ is calculated from equation 2.16 as

$$\hat{T}_{11} = \frac{0.0046 \times 160 \times 1.45 \times 200}{2} = 107$$

The values for the other balancing factors are $A_2 = 0.00545$, $A_3 = 0.0045$, $B_2 = 0.735$, and $B_3 = 1.25$.

Notice in the predicted flow table that there is a constraint on total outflows and on total inflows. Generally, when increasing constraints are placed on the gravity model, the predicted interactions, not surprisingly, tend to become more accurate.

This chapter has outlined the basic forms of the gravity model, how the different models are related, and how they perform in a forecasting situation. A similar analysis can be found in Senior (1979). More advanced relationships between the different members of the gravity model family can be found in Fotheringham and Dignan (1984).

3. GRAVITY MODEL APPROACHES TO MARKET ANALYSIS

The gravity model has been widely applied in the marketing field. Its use as an estimation method in service area analysis is well established. However, much of this work has been limited to defining the trade area or hinterland market for a retailing center. Market area estimation is only one aspect of the market analysis use to which spatial interaction models can be put, and is a rather limited view of the complex area of marketing.

As early as 1958, a series of fundamental marketing issues were identified by Homer Hoyt, all of which must be effectively integrated for estimating decision-making in marketing. These issues included the following: (1) estimation of demand; (2) identification of a retail center's trade area or market hinterlands; (3) determination of the appropriate size of a retail center; and (4) estimation of optimal location decisions in terms of market access or revenue maximization. Furthermore, (5) specialized and local market characteristics should be identified, and (6) the impact of existing demographic patterns should be forecast in terms of future market impacts. The gravity model can easily be adapted to examine each of these issues. Below we highlight some of the marketing uses of this multifaceted technique. We will first review the evolution of gravity modeling in marketing, and will then integrate our descriptive review with interpretive considerations that are consolidated as applications in Chapter 5.

The individuals whose work dominate the literature in the application of the gravity model to retailing and marketing are Reilly (1929, modified by Converse 1949) and Huff (1959). Both of these researchers introduced simple, practical, and useful applications of the gravity concept to describing, explaining, and predicting market behavior. Reilly examined trade areas, market sheds, or service area (retailing) hinterlands focusing on the aggregate competitive effects of alternative markets. Huff's applications focus on the response of consumers, shoppers, or patrons to the location of retail outlets and shopping centers.

Market Area Boundaries

Since the 1930s, Reilly's general model has become so widely known and so commonly used that it is often referred to as a law, Reilly's Law of Retail Gravitation (Batty 1978). The two gravity principles of scale and distance apply. In terms of scale, the larger the city the more retail trade it is expected to draw from towns in the surrounding area. In terms of distance, a city is expected to draw more trade from nearby towns than from more distant ones. Following the early gravity model, Reilly (1929) suggested,

> A city will attract retail trade from a town in its surrounding territory, in direct proportion to the population size of the city and in inverse proportion to the square of the distance from the city.

The simplest formal representation of this is that city i will be attractive for retailing, A, to individuals in city j in direct proportion to the population P at i and inverse proportion to the square of the distance from city i to city j, d_{ij}^2.

$$A_i = \frac{P_i}{d_{ij}^2} \qquad [3.1]$$

This might be seen as a one dimensional potential interaction flow from j to i similar to that outlined in Chapter 1. However, Reilly expanded the analysis by examining the competition between centers for a market. Using this simple base he developed a method for estimating the flow of consumers and/or expenditures from a market hinterland or area to competing market centers. Next, he extended this method into a market boundary identification technique.

In his terms, the competing-centers method allocates consumer flows from the intermediate area between the centers in proportion to the attractiveness of the centers, measured by population size, and inversely with respect to the square of distance. A city i with a population P_i and a city j with a population P_j would attract population from an intermediate center x at distance d_{ix} and d_{jx} in proportion to the attractiveness values of A_i and A_j. This proportion is a direct ratio of $A_i{:}A_j$ where

$$A_i = \frac{P_i}{d_{ix}^2} \text{ and } A_j = \frac{P_j}{d_{jx}^2} \qquad [3.2]$$

An example makes this clear. Dallas and Oklahoma City are approximately 200 miles apart with a metropolitan population of 2.5 million and 1 million, respectively. Wichita Falls is an intermediate city of 350,000, 100 miles from Dallas and 140 miles from Oklahoma City. If Dallas is city i and Oklahoma City is city j, then the market attraction of these centers for expenditures which flow from Wichita Falls to Dallas is

$$A_i = \frac{P_i}{d_{ix}^2} \quad \frac{2{,}500{,}000}{100^2} = \frac{2{,}500{,}000}{10{,}000} = 250.00$$

and from Wichita Falls to Oklahoma City is

$$A_j = \frac{P_i}{d_{jx}^2} \quad \frac{1{,}000{,}000}{140^2} = \frac{1{,}000{,}000}{19{,}600} = 51.02$$

which gives the ratio

$A_i{:}A_j$ = 250.00:51.02 = 250 ÷ (250+51.02) or 83.05 percent: 51.02 ÷ (250 + 51.02) or 16.95 percent.

The trade flow or market share of expenditures from Wichita Falls to Dallas and Oklahoma City will be divided such that 83.05 percent goes to Dallas and 16.95 percent goes to Oklahoma City.

Reilly used this information to determine the distance at which the attraction power of the two centers would be equally balanced. The distance that divides the market flows equally (50 percent:50 percent) was defined as the breaking point or market area boundary. In reality, it is not the market area boundary but rather the point at which the dominance of one market center shifts to another. Consumers at this boundary should be indifferent as to which market center they patronize. It is the point at which the attraction of each market center is equal, $A_i = A_j$, or

$$\frac{P_i}{d_{ix}^2} = \frac{P_j}{d_{jx}^2} \qquad [3.3]$$

Using equation 3.3 and the fact that $d_{ix} = d_{ij} - d_{jx}$ we can rewrite this as

$$d_{ix} = \frac{d_{ij}}{1 + \sqrt{P_j/P_i}} \qquad [3.4]$$

Note that in the derivation of equation 3.4 we assumed the exponents $\lambda = 1$, $\alpha = 1$, and $\beta = 2$ were appropriate. The general form of the model, without specifying exact values for the exponents, is

$$d_{ix} = \frac{d_{ij}}{1 + \left(\frac{P_j^{\alpha}}{P_i^{\lambda}}\right)^{1/\beta}} \qquad [3.5]$$

For the Dallas-Oklahoma City case, equation 3.4 is solved as

$$\frac{200}{1 + \sqrt{1{,}000{,}000/2{,}500{,}000}} = \frac{200}{1 + \sqrt{.4}} = \frac{200}{1.6325} = 122.5$$

This means the distance of the breaking point from Dallas, or the dominance equilibrium, d_{ix}, is 122.5 miles. Obviously from this we can deduce that the breaking point is 77.5 miles (200-122.5) from Oklahoma City. These relationships are shown in Figure 3.1.

In Figure 3.2 we incorporate basic elements of Figure 3.1 to develop the trade-off relationships between market center attraction, distance, and market share dominance for the Dallas-Oklahoma City case. We can see the breaking point or market boundary is where the attractions of the two centers are equal. Due to the greater population size of Dallas, its market dominance extends further than Oklahoma City's market dominance. Although Wichita Falls is not on the boundary line between the two centers and falls within Dallas's market dominance area, this only means that a majority of Wichita Falls expenditures, 83.05 percent, will flow to Dallas. The curves that relate market share dominance to each center can be interpreted not only as shares but also as probabilities. The probability that consumers at 122.5 miles from Dallas and 77.5 miles from Oklahoma City (at the breaking point) will choose Dallas over Oklahoma City is .50. Similarly, at any distance between Dallas and Oklahoma City we can read off these probabilities by using the curves that represent attraction proportions. This is an important characteristic, not appreciated by Reilly, but

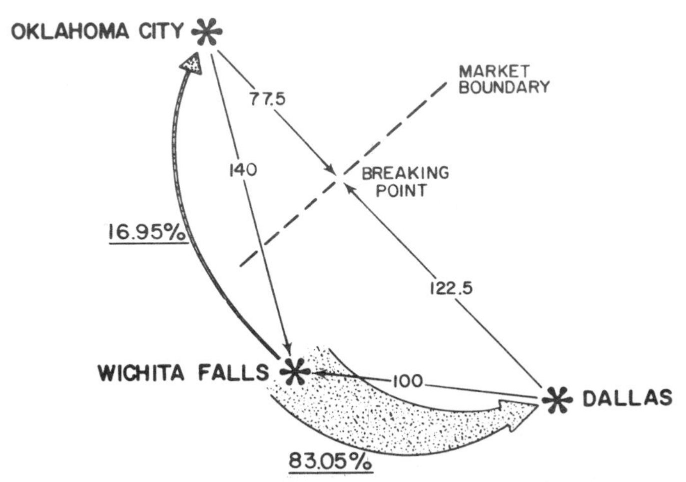

NOTE: This illustrates the distance configuration between Dallas and Oklahoma City and the market hinterland boundary between centers together with impact of market attraction on Wichita Falls expenditure flows to the next largest centers.

Figure 3.1 Dallas, Oklahoma City, Wichita Falls Hinterland Configuration

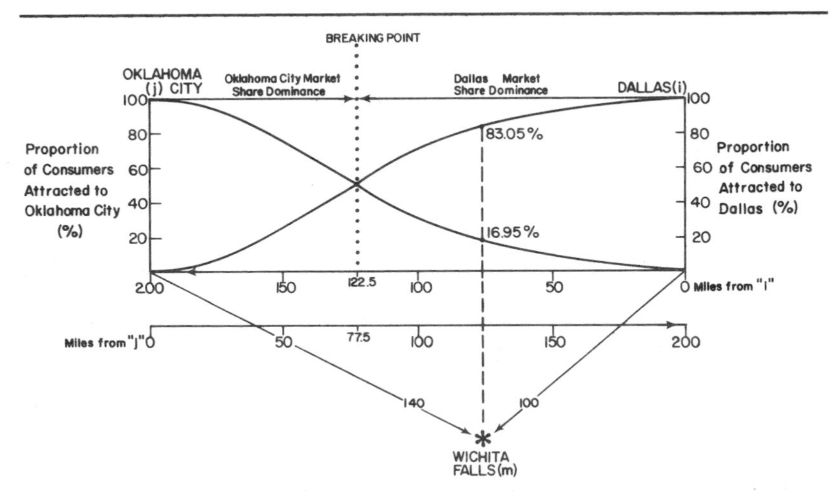

NOTE: This illustrates the market share distribution between Dallas and Oklahoma City and the allocation of Wichita Falls expenditures flowing to the next larger centers.

Figure 3.2 Market Share Diagram

widely used by others in extending the retail model in both theoretical interpretation and practice.

The results presented for the Dallas-Oklahoma case are simplified with only two competing centers and one intermediate center. A unidirectional system was examined. Obviously these results will vary by direction in response to the attraction level of market centers. A multicenter, multidirectional example of trade area competition for market hinterland dominance using this breaking point or equilibrium identification approach is given in Chapter 5 as Example 2. This approach using a seven-city system should provide the reader with the ability to generalize Reilly's model to most real-world situations. This is an application of the total flow constrained gravity model discussed in Chapter 2.

The application discussed here is a simplification of reality in several ways. Reilly's law only provides information about general patterns. Market areas are not permanently fixed in time or space and often have specialized characteristics. Furthermore, the effect of distance is likely to vary by region, mode of transportation, type of retailing, and consumer incomes, as well as by shopping frequency. This could be indicated by variations in the distance exponent β. One systematic variation found in these exponents reflects price and frequency of purchase characteristics. Larger exponents, and hence local patterns of shorter travel distances, are associated with the purchase of convenience goods while smaller exponents, and hence regional patterns and longer travel distances, are associated with the purchase of consumer durables. However, Reilly's approach, although limited in behavioral content and deterministic in original construction, provided the seeds for further development in market analysis.

Demand and Market Thresholds

When planning a new retail service or center, the trade area boundary helps to focus analysis on the threshold demand needed to cover costs and generate a profit from supplying a new service. Planning for a new service must take into account that the demand for that service is often distributed over a county, metropolitan region, or trade area. Hence, the further from the point at which the service is offered, the higher the price due to transportation and time costs; and the higher the price, the lower the demand. Lösch's (1954) demand cone, which describes the decline in demand as the distance increases from the point where a service is offered, incorporates the "friction of distance" gravity model concept into higher prices (see Figure 3.3). This is one element of demand estimation and is discussed in the review of central place concepts by L.J. King (1984) in another book in this series.

In Example 1 in Chapter 5 we estimate the latent or unexpressed demand for a service—air transportation—in assessing the feasibility of of-

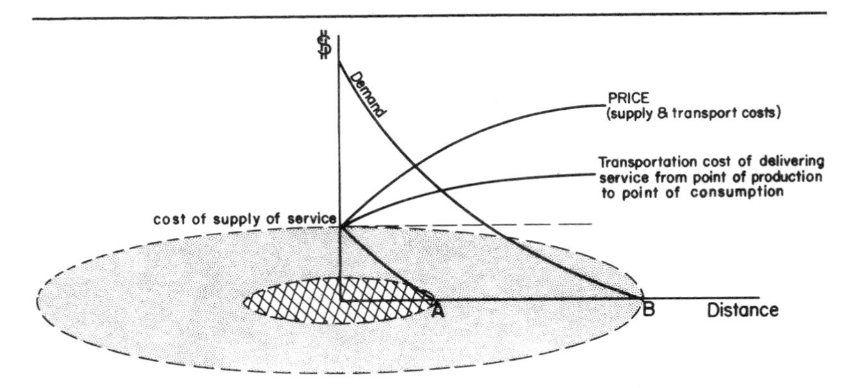

NOTE: This illustrates the impact of transport costs on product or service price as one increases the distance from the point at which the good or service is offered to the point where the good or service is consumed. Assuming a normal good, demand goes to zero when the price gets large (at point B). A minimum market is needed to cover the costs of producing and offering a good or service. This minimum market is the threshold for survival (denoted by the market up to point A).

Figure 3.3 Lösch's Demand Cone

fering a new airline route. If we convert our costs for offering a service to a minimum number of customers or users needed to make the service break even, then this establishes our minimum threshold or hurdle level. If estimates show that we are above this level we can institute the new service; if not, we know the flow of users will not cover costs. In the private decision case, this tells us how much demand will have to grow in order for the "go-ahead" decision to be profitable. In the public sector case, it tells us that if for equity, social, or humanitarian reasons, we should go ahead with offering the service then we will fall short of covering our costs by a certain amount. This short-fall is the level of public subsidy that will be required should one decide to offer the service.

The air transportation example is an assessment of the latent demand for transportation between two points based on population size of the points (cities) and the distance apart. The decision is made to go ahead and offer the new service because the model predicts higher user rates than would be needed to cover costs. The size of that surplus is an estimate of potential profits from the decision to offer the new air service link.

The Operational Retail Model

Huff's contributions to retail modeling are extremely fundamental and hence important. From a theoretical perspective, he recast the gravity model from a deterministic to a probabilistic perspective (1963a, 1963b) and refocussed attention from aggregate outcome to consumer behavior (1959,

1960, 1961). In his Los Angeles study, Huff (1962) essentially operationalized Luce's (1959) individual choice axiom to retail spatial behavior and in so doing laid the groundwork for interpreting the empirical regularities of the gravity model as outcomes of human decision making. This theoretical breakthrough allowed for major operational advances. From an operational perspective Huff introduced a practical approach to defining the "attraction" of a center as the amount of floor space rather than simply population (1964). This opened up the intepretation of attractiveness and allowed it not only to be determined by a number of variables (e.g., number of functions, size of parking capacity, etc.) but also allowed attractiveness to be treated as an independent variable that could be estimated in its own right. Another major operational consideration was that Huff fitted the exponent for distance in trip-making behavior to particular circumstances. Finally, he introduced a balancing term that constrained the sum of individual or zonal travel or sales to fit within an overall travel or sales limit.

With respect to the attractiveness or drawing power of a market center, Huff's use of retail floor space has been widely adopted and adapted to include other important characteristics (1966). Most important though, this operationalization "demystified" the idea of drawing power or attraction and allowed its direct estimation by focusing on the weight or exponent associated with it. Nakanishi and Cooper (1974) were particularly effective at utilizing Huff's probabilistic choice framework and operational attractiveness perspective to develop a linearization procedure for direct estimates of attractiveness. It is the balancing term of this production-constrained model that allows for accurate estimation of attractiveness. Similarly, it was this balancing term that Lakshmanan and Hansen (1965) used to make sure that their cash flow or expenditure variable allocated to various retail centers would be within an overall expenditure limitation. From this it follows that the level of overall expenditures within a region is the limiting characteristic in allocating customers to retail centers. Indirectly, then, we have information about the sizing of proposed new retail centers in a region. Such centers, if they are to survive, will be limited to the size of the market in terms of retail expenditures. Using this basic logic it is possible to size explicitly a new center for the unmet demand in a specific market.

The advantages of sizing new retail centers appropriately should be obvious (Young, 1975). If they are sized too large then the returns per square foot of floor space are too low and profitability of an entire project may be in jeopardy. If it is sized too small then unused surplus demand may attract competition. The assumption is that the new retail center is competing in a closed market and that it cannot by its presence, increase demand either by increasing the region market area or by increasing existing consumer's expenditure levels. This is probably a good conservative assumption for small new retail entrants, but may be restrictive for the largest shopping centers that hope to expand service based on increasing the region's overall trade area. Further, this makes no assumption about the actual store level competitiveness of various outlets but sizes a center relative to a homogeneous demand. From another perspective this approach allows a manager to set a square foot sales goal for a center relative to its location and overall sales potential. Sales above this level must be attributed not to location or market size but rather to sales force skill and competence. Public sector considerations relate to the development of appropriate zoning in a local area to assure that the amount of retailing space required by a residential population be set aside for that purpose. It was this basic consideration that stimulated Huff's early probabilistic applications of intraurban retail modeling in the situation of rapid residential population expansion in Los Angeles of the late 1950s and early 1960s. An example of the optimal sizing of a shopping development in a four-town unified market setting is given in Example 5 of Chapter 5

The Lowry Model

Lowry (1964) developed an urban land use model that was built around two gravity model structures with residential and retail service feedbacks. The model utilizes overall constraint characteristics as balancing tools and introduces an interactive, interdependent framework for gravity models. An attraction constrained model relates the distribution of population to residential zones from an initial distribution of basic employment by zone. People are allocated via the model from job locations to residential locations. This is the household sector. The household sector generates a demand for services that are met by a second gravity model—a production constrained model—that allocates retail service to this demand. The feedback occurs by retail services generating employment that then has to be allocated back to residential locations. The iteration continues as this new distribution of residents generates a demand that has to be met by a new round of retail service distribution. This interactive process continues until some arbitrarily small number of residential and service reallocations are demanded. A general outline of the process is given in Figure 3.4.

The Lowry model was the first of a series of dynamic models that linked urban structure to basic employment, retail/service demand, and residen-

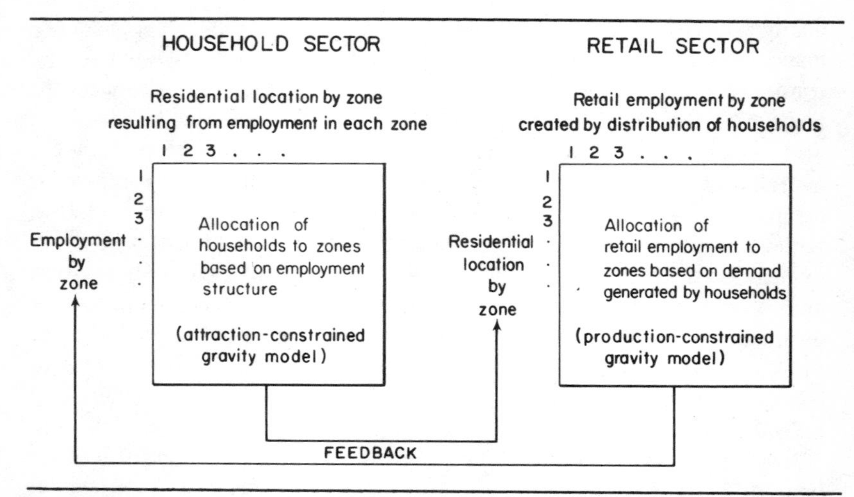

Figure 3.4 Lowry Model Outline

tial location (Lowry, 1967). Although there are a number of variations on this basic theme (Goldner, 1971), and some substantial improvement, many still incorporate Lowry's residential-retail components in both intraurban and regional growth models (Haynes 1968; Putman 1979; Foot 1981; Fotheringham 1984b).

The Retail/Service Model

The use of a retail service model is presented in Example 6 of Chapter 5. This model incorporates many of the issues raised up to this point and deals with both market/service access and revenue maximization. The former is important not only to retailing but also to public sector decision making. With respect to the latter, services are offered to a population as a response to demands or needs often expressed in the political process. Trying to maximize the access of a potential population—the handicapped, the elderly, or the poor—is of vital importance for the maintenance and continuance of these services. Use levels are always an important component of arguments related to budgetary maintenance or expansion. Further, it is suggested by many that high use levels may generate some economies of scale, albeit often limited, even in service delivery systems.

From a private perspective, professional services (Morrill 1959) and retailing have a lot in common in terms of their needs to access this spatially dispersed market. Clearly, other characteristics related directly to the attributes of the service delivery location (such as parking, building characteristics, safety, security, and neighborhood elements) also affect the use level of centers offering public, private, and retail service, but access is a particularly important choice criteria in large complex systems such as U.S. metropolitan regions.

The concept of revenue maximization is related primarily to retail considerations. However, it also affects professional services where "drop-in" service or high competition and high minimum service standards are externally established; thus, for professional services location becomes an area of variability for choice criteria. Revenue maximization may also be an important consideration, over and above access, in public sector services where user charges are a major component of the operating budget such as automobile license branches, municipal recreation (golf courses, etc.), and public museums and art galleries. Although retailing correctly dominates our example, these other application opportunities should not be overlooked.

Two other applications of gravity models that can be used in important ways in market analysis are found in Chapter 5 (Examples 3 and 4). In Example 4 the new service being offered is located at a fixed destination—a state university or the location of a professional meeting. However, it could also be the location of a shopping center or retail outlet. In this application we are not asking about relocation or establishment of a new outlet but rather we are trying to understand special characteristics of our existing market pattern more effectively. In this case we are interested not so much in the correct estimates generated by the model but in the distribution of incorrect estimations. After controlling for location, why do we still draw more or fewer people from a zone—a residential area or a state—than expected? Provided we are using a reliable model, these residuals—under- and over-predictions—may contain important information patterns. Are the determinants of these residuals income related or knowledge related, or do they have other market implications? Once identified we can then develop strategies to address these areas better (e.g., advertising, cheaper transportation, shift in price patterns of goods offered, etc.).

Example 3 in Chapter 5 is not unrelated to the issues above, but is more of a strategic planning concern. The example deals with voting patterns and migration; but we could translate voting patterns into expenditures between competing retail chains, and migration could produce changes in certain demographic characteristics such as age, sex ratio, or occupational structure that affect purchase choices. If we can forecast a change in one of these demographic characteristics, we can anticipate changes in expenditure patterns. Such changes could occur in the level of expenditure or in the purchase pattern of goods. Retail outlets are large capital investments often with twenty- to thirty-year time horizons. Shopping complexes are even larger investments with even greater time commitments. These capital investments are fixed in space, and even after depreciation, we would still like to maximize their revenue generating capacity. Being able to define market sheds or trade areas allows us to forecast special demographic characteristics for those areas and estimate related market changes. These strategic considerations can be done economically, either in metropolitan submarkets for a service of competing retailers or nationwide for a given retail chain.

Trends

We have only begun to supply you with some of the marketing applications in which the gravity model may be effecively used. Since the mid-1960s the marketing literature has exploded in both the volume and range of applications. The emphasis in retailing location has focused on the attributes of attraction points and the service characteristic of individual outlets. These are the areas over which many retailers feel they have greatest control. Although the location-relocation issues continue to be important, strategic planning and analysis of existing market patterns may also provide important knowledge for market leverage.

4. ORIGIN- AND DESTINATION-SPECIFIC GRAVITY MODELS

The general gravity model formulation of equation 1.5 has been used to generate four specific model forms: the total flow constrained gravity model in equation 2.3; the production-constrained gravity model in equation 2.7; the attraction-constrained gravity model in equation 2.12; and the doubly constrained gravity model in equation 2.15. Each of these models is used to represent a system of flows or interactions between many origins and many destinations such as that depicted in Figure 4.1, part a. Each model, however, can also be applied to flow systems such as that of Figure 4.1, part b, where there are flows from only one origin, or to flow systems such as that of part c in Figure 4.1 where there are flows into only one destination. In the former situation, the gravity models are described as origin-specific because the results are specific to one origin; in the latter situation, the models are described as destination-specific because the results are specific to one destination. It is often interesting and enlightening to compare model performance and parameter estimates between origins and/or destinations. Information can then be obtained on behavior specific to each origin or destination.

Origin-Specific Gravity Models

The origin-specific version of the total flow constrained gravity model is

$$T_{ij} = k_i w_j^{\alpha_i} d_{ij}^{\beta_i} \qquad [4.1]$$

where the variables are defined as before and the parameters k_i, α_i and β_i are specific to origin i. The parameters α_i and β_i reflect the perception of destination attractiveness and distance as determinants of interaction by the residents of origin i. The balance of total flows is ensured by k_i. As $v_i^{\lambda_i}$ is a constant in the model (there is only one origin), it is subsumed in k_i and can be ignored.

The origin-specific version of the production-constrained gravity model is

$$T_{ij} = A_i O_i w_j^{\alpha_i} d_{ij}^{\beta_i} \qquad [4.2]$$

where

$$A_i = [\sum_j w_j^{\alpha_i} d_{ij}^{\beta_i}]^{-1} \qquad [4.3]$$

There is no origin-specific version of the attraction-constrained gravity model because such a model would trivially constrain each predicted flow to equal the equivalent actual flow.

The origin-specific version of the doubly constrained gravity model is

$$T_{ij} = A_i O_i B_j D_j d_{ij}^{\beta_i} \qquad [4.4]$$

where

$$A_i = [\sum_j B_j D_j d_{ij}^{\beta_i}]^{-1} \qquad [4.5]$$

and

$$B_j = [\sum_i A_i O_i d_{ij}^{\beta_i}]^{-1} \qquad [4.6]$$

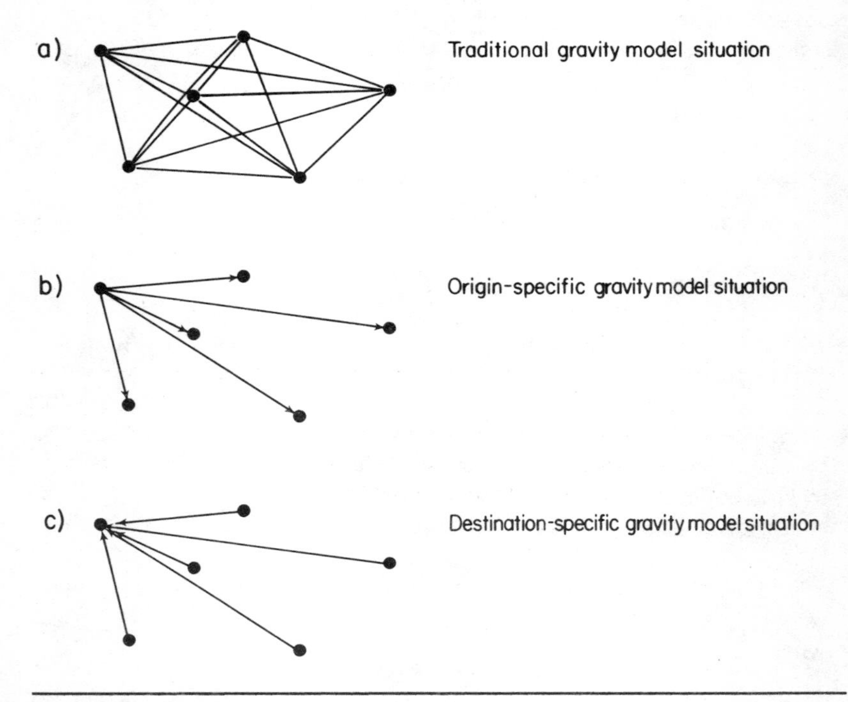

Figure 4.1 Interaction Systems

Note that in the summation of A_i, β_i will be constant because the summation is over j; but in the summation of B_j, β_i will vary in each element of the summation because the summation is over i.

Destination-Specific Gravity Models

The destination-specific version of the total flow constrained gravity model is

$$T_{ij} = k_j v_i^{\lambda_j} d_{ij}^{\beta_j} \qquad [4.7]$$

where k_j, λ_j and β_j are parameters specific to destination j. As $w_j^{\alpha_j}$ is a constant in the model (there is only one destination) it is subsumed in k_j.

There is no destination-specific version of the production-constrained gravity model since such a model would trivially constrain each predicted flow to equal the equivalent actual flow.

The destination-specific version of the attraction-constrained gravity model is

$$T_{ij} = B_j D_j v_i^{\lambda j} d_{ij}^{\beta j} \quad [4.8]$$

where

$$B_j = [\sum_i v_i^{\lambda j} d_{ij}^{\beta j}]^{-1} \quad [4.9]$$

The destination-specific version of the doubly constrained gravity model is

$$T_{ij} = A_i O_i B_j D_j d_{ij}^{\beta j} \quad [4.10]$$

where

$$A_i = [\sum_j B_j D_j d_{ij}^{\beta j}]^{-1} \quad [4.11]$$

and

$$B_j = [\sum_i A_i O_i d_{ij}^{\beta j}]^{-1} \quad [4.12]$$

In this case, β_j varies in the summation of A_i, but is a constant in the summation for B_j.

Applications of Origin- and Destination-Specific Gravity Models

Consider an actual interaction system consisting of flows between many origins and many destinations. When a model is calibrated in such a system, a maximum of one parameter is estimated for each variable. For instance, one distance-decay parameter is estimated for the whole system. Obviously, this is an "average" of the individual distance-decay parameters for each origin or for each destination and as such suffers the same problems as an arithmetic average does in trying to describe a set of numbers—it can hide a great deal of information. Much more information can be gained on the system under investigation if origin-specific or destination-specific parameters are estimated. The parameter estimates can then be compared against one another or against some average value and conclusions can be drawn regarding interaction behavior from each origin or to each destination. For example, suppose that for two origins, A and B, the estimated distance-decay parameters are –0.5 and –2.0, respectively. We might con-

clude on the basis of this information that distance is perceived to be a greater deterrent to interaction by the residents of place B than by the residents of place A. This prompts the interesting question of why this might be so. A similar research question would be prompted if the distance-decay parameters were destination-specific instead of being origin-specific. Thus, a major justification for the use of origin-specific and destination-specific models is that they provide a great deal more information than do traditional models.

Origin-specific gravity models have probably been more widely employed than destination-specific models. Information pertaining to origins is perhaps more intuitively appealing and more useful than information pertaining to destinations because the former is based on a more homogeneous set of interactions. Migrants or commuters, for example, starting out from the same origin are more likely to have characteristics in common with one another than are migrants or commuters choosing the same destination. To give a more concrete example, in the United States it is intuitively acceptable to make the statement that people *from* the Midwest perceive distance to be a greater deterrent to interaction than do people *from* the West Coast. However, it is much less intuitively acceptable to make the statement that people who move *to* the Midwest perceive distance to be a greater deterrent to interaction than do people who move *to* the West Coast. The former is obviously the kind of statement that could be made by calibrating origin-specific gravity models while the latter is the kind of statement that could be made by calibrating destination-specific gravity models.

Origin-specific and destination-specific gravity models have been used to study interactions in many situations including international trade flows (Linneman 1966), airline passenger interactions (Fotheringham 1981), migration (Greenwood and Sweetland 1972; Stillwell 1978), freight flows (Chisholm and O'Sullivan 1973); telephone calls (Leinbach 1973); and the transmission of knowledge (Gould 1975).

A Cautionary Note Concerning Spatial Structure

In almost all studies of origin-specific distance-decay parameters, a noticeable spatial trend in estimated parameter values has been reported: Estimated distance-decay parameters tend to be less negative for accessible (or central) origins than for inaccessible (or peripheral) origins. This trend, for example, is reported in all of the above studies. Taken at face value the trend would suggest that residents of more accessible origins perceive distance as less of a deterrent than do residents of less accessible origins (Haynes, Poston, and Schnirring 1973). Fotheringham (1981), however, suggests that this would be a misleading interpretation and that the trend in parameter estimates reflects an error in the gravity model formulation—a relaionship between interaction patterns and spatial structure exists but is not included in the gravity model formulation. Spatial structureis a term used to denote the spatial arrangement of origins and destinations in the interaction system under investigation. The relationship between interaction patterns and spatial structure has also been referred to as the "map pattern effect" (Johnston 1975; Cliff, Martin, and Ord 1975; Curry, Griffith, and Sheppard 1975).

Fotheringham describes five pieces of evidence derived from origin-specific distance-decay parameter estimates that suggest the misspecification of the gravity model formulation:

(1) the strong relationship between the relative location of origins and their associated distance-decay parameter estimates;
(2) the heterogenity of behavior suggested by the large variation in parameter estimates in what are often expected to be relatively homogeneous systems;
(3) the occasional reporting of positive distance-decay parameters contrary to all intuitive understanding of interaction behavior. They seemingly indicate that as distance from an origin increases, interaction increases, *ceteris paribus*.
(4) The relationship between β_i and mean trip length is frequently reported as being negative when we would generally expect it to be positive. That is, as the mean trip length for a given origin increases (and hence as people travel longer distances), then β_i should become less negative. That such a relationship is not usually found is probably a result of the relationship described in (1) as mean trip length is likely to be shorter for accessible origins than for inaccessible origins.
(5) Occasionally, values of particular distance-decay parameter estimates are at odds with our intuition. Fotheringham (1981) gives an example in his U.S. airline passenger study where the parameter estimates were much less negative for cities such as Albany, Syracuse, and Utica than for cities such as Los Angeles, San Francisco, and Las Vegas. It would appear that the former group of cities contain "jet setters" while the latter group contain a relatively more parochial group of people!

Competing Destinations Models

It is one thing to say in very general terms that spatial structure affects interaction patterns, and another to specify exactly what it is about spatial structure that is not already included in the gravity model formulation and that determines interactions. The latter was a topic of fairly intense research throughout the 1970s (Curry 1972; Johnston 1973; and Fotheringham and Webber 1980). This research generated some interesting ideas but was generally inconclusive. In a recent series of papers, however, Fotheringham

(1983a, 1983b, 1984a) supports the contention that the "missing variable" from the general gravity model formulation is the competition each destination faces from all other destinations. Destinations are viewed as competing with each other for interactions and when a variable measuring such competition is included in the gravity framework, the resulting interaction models are known as competing destinations models.

Three justifications for the addition of a variable describing the competition between variables can be made:

(1) Consider parts a and b of Figure 4.2. In each system there are seven destinations of equal attractiveness and at equal distance from a single origin. In part a of Figure 4.2 the destinations are spaced at equal intervals from one another, whereas in part b some destinations are clustered and some are isolated. Intuitively, we might expect that the different arrangements of the destinations would produce some variation in the interaction patterns. For some types of interactions, such as non-grocery shopping, a destination may attract a greater number of customers if it is located in close proximity to other possible destinations (as epitomized by the shopping mall concept). For other types of interactions, such as grocery shopping, a destination that is relatively isolated from other possible destinations may be able to capture a local market, and so attract more customers, because of the lack of competition.

(2) Consider interactions being the result of a two-stage decision-making process. People first choose a macrodestination and then choose a microdestination within that macrodestination. For example, an unemployed worker in Michigan searching for work may first decide to move to the South (the macrodestination). He or she then selects a particular location within the South (the microdestination). Similarly, an individual in Indianapolis may decide to vacation in New England, Florida, California or the Pacific Northwest. His or her first locational decision is to choose one of these broad regions; the second locational decision is to choose a specific destination within the chosen region. If such hierarchical decision-making behavior does occur in reality, then the more microdestinations there are within a macrodestination, the greater will be the competition between microdestinations and the fewer interactions will terminate at any particular microdestination.

(3) Everything else being equal, places that are relatively isolated tend to gain more recognition than places that are clustered and perhaps overshadowed by larger neighbors (see Gould and White 1974). For example, if a person is asked to list fifty U.S. cities he or she is more likely to mention Charleston, South Carolina, or Albuquerque, New Mexico, than Utica, New York, or Binghamton, New York, despite the similarity in the populations of these cities. The greater recognition of centers in relative isolation could produce greater than anticipated interaction to such places.

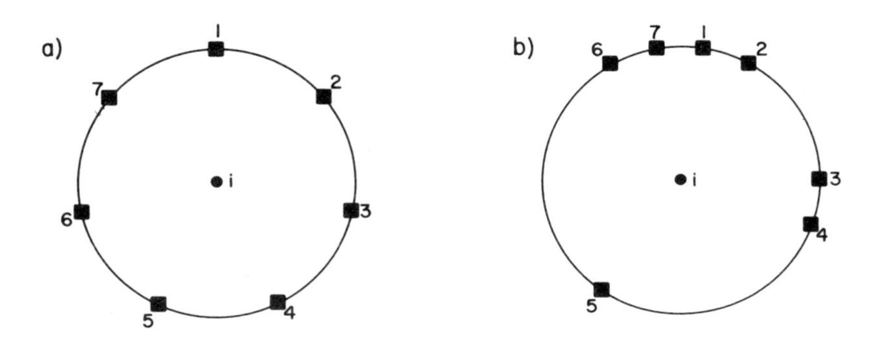

Figure 4.2 Two Spatial Arrangements of Destinations

The inclusion of a variable measuring destination competition within the gravity model framework is straightforward. One possible measure of destination competition is the potential accessibility measure (Hansen 1959):

$$c_j = \sum_{\substack{k \\ k \neq j}} w_k^{\alpha} / d_{kj}^{\beta} \qquad [4.13]$$

where c_j represents the accessibility of destination j to all other destinations; w_k represents the attractiveness of destination k; d_{kj} represents the separation between k and j; and α and β are defined as in the gravity model. Often, they are set equal to 1.0 in the accessibility formulation.

The general competing destinations model framework is

$$T_{ij} = f(\underline{V}_i, \underline{W}_j, \underline{C}_j, \underline{S}_{ij}) \qquad [4.14]$$

where $\underline{C}_j$ represent a vector of competition variables. A total flow constrained competing destinations model derived from the general formula is

$$T_{ij} = k v_i^{\lambda} w_j^{\alpha} c_j^{\delta} d_{ij}^{\beta} \qquad [4.15]$$

where δ is a parameter relating interactions to destination competition. A production-constrained competing destinations model is

$$T_{ij} = A_i O_i w_j^{\alpha} c_j^{\delta} d_{ij}^{\beta} \qquad [4.16]$$

where

$$A_i = [\sum_j w_j^{\alpha} c_j^{\delta} d_{ij}^{\beta}]^{-1} \qquad [4.17]$$

We will not describe the attraction-constrained and doubly constrained versions of the competing destinations models as it has been demonstrated that they are very similar to the equivalent gravity models in many situations (Fotheringham 1983a). The reason for this is that the definition of the balancing factor B_j in such models is similar to the definition of c_j in equation 4.13 and can act in the same way to model interaction patterns.

The spatial structure effect described very briefly above is one of the more intriguing and more complex problems in geography: How, and to what extent, is behavior determined by the spatial arrangement of places. In the case of interaction modeling, it appears that we are on the way to answering this question but it is a question that should be asked in all situations where human behavior is being modeled. With reference to the section on gravity modeling and retailing, it is possible that advances in retail modeling can be achieved by substituting a production-constrained competing destinations model for a production-constrained gravity model (Fotheringham 1984b).

5. USES AND EXAMPLES OF THE GRAVITY MODEL

In this chapter we describe some typical uses of the gravity model in real-world situations. The examples are described in approximate order of increasing complexity; and while more detail on particular examples is given at various points within the preceding chapters, this chapter can be read independently by a reader wanting to know whether the gravity modeling technique is worth pursuing in his or her research. The uses described are as follows:

(1) planning a new transportation service (or a new road);
(2) defining retail shopping boundaries;
(3) understanding the effect of migration on voting patterns (or on consumption patterns);
(4) analyzing university enrollments by state;
(5) determining the optimal size of a shopping development;
(6) locating a facility for maximum patronage.

These uses are not exhaustive but are meant to give an idea of the situations in which the gravity modeling framework can be applied. The reader is encouraged to think of alternative uses.

Example 1: Planning a New Service

Forecasts of interactions are often used as a basis for determining whether a new transit service will be profitable or whether a new road is needed between two centers. Here we examine airline passenger interactions and the feasibility of a new airline route.

Suppose a total flow constrained gravity model is calibrated using data on air passenger traffic per year between U.S. cities. The calibrated model is

$$\hat{T}_{ij} = (9.128 \times 10^{-16}) \times (v_i^{2.1}) \times (w_j^{2.0}) \times x\ (d_{ij}^{-0.56}) \quad [5.1]$$

where $\hat{T}_{ij}$ represents the predicted interaction between i and j; v_i represents the population of origin i; w_j represents the population of destination j; and d_{ij} represents the distance between i and j. The value 9.128×10^{-16} is the estimate of k, the constant that ensures that the total predicted flow equals the total actual flow. The notation means that 9.128 is divided by 10 sixteen times so k is obviously very small (in fact, it is 0.0000000000000009128).

Assume that a local airline company wants to know if it is profitable to initiate a regular flight between two cities, x and y. City x has a population of 100,000 and city y's population is 300,000. The distance between them is 500 miles (these units are the same as those used to calibrate the model). No airline service presently exists between these two cities and the airline company needs to be assured of at least 50,000 people per year travelling in *each* direction and a total passenger volume of at least 120,000. On the basis of the calibrated gravity model, would you recommend to the airline that it initiate a regular service? We can answer this question by putting the appropriate data into the model and obtaining estimates of interaction volumes:

$$\hat{T}_{xy} = (9.128 \times 10^{-16}) \times (100{,}000^{2.1}) \times (300{,}000^{2.0}) \times (500^{-0.56}) = 80{,}021$$

and

$$\hat{T}_{yx} = (9.128 \times 10^{-16}) \times (300{,}000^{2.1}) \times (100{,}000^{2.0}) \times (500^{-0.56}) = 89{,}313$$

so that $\hat{T}_{xy} + \hat{T}_{yx} = 169{,}334$.

Consequently, our predictions suggest that the criteria for the new air service to be profitable are likely to be met. The assumptions are being

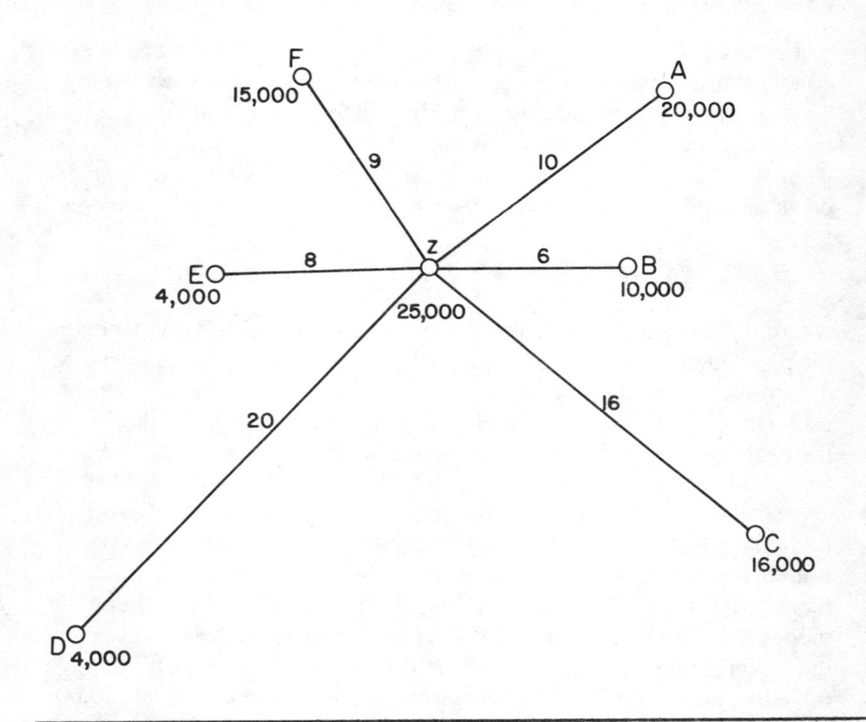

Figure 5.1 A Spatial System

made, of course, that the model is accurate and that the behavior of the people in cities x and y is the same as that of people in the cities in which the model was calibrated.

Example 2: Defining Retail Shopping Boundaries

Suppose that we have calibrated the total flow constrained gravity model with data on shopping patterns between the cities in Figure 5.1. Origin propulsiveness, v_i, and destination attractiveness, w_j, are both measured by population size; hence, we use the notation P_i and P_j in this example. Distances are used to measure spatial separation and are denoted by d_{ij}. Our task in this instance is to determine the market area around city Z. The market area boundary between two centers is the point at which people are indifferent as to which of the two centers they shop in.

From Riley's Law of Retail Gravitation (see Chapter 3), the location of the market boundary between i and j is given by the distance d_{ix} where

$$d_{ix} = \frac{d_{ij}}{1 + \left(\frac{P_j^{\alpha}}{P_i^{\lambda}}\right)^{1/\beta}} \qquad [5.2]$$

The market boundary around city Z (which is our origin i in this example) can then be drawn by first calculating the location of the boundary between Z and each of the other cities. For each competing city this boundary lies at point x, which is at a distance d_{ZX} from Z. Perpendicular lines are then drawn at x across the straight line joining Z with its competitor. These perpendicular lines are then extrapolated to produce a market area for place Z. In this example, assume $\lambda = \alpha = 1.0$ and $\beta = 2.0$. Then, for center A,

$$d_{ZX} = \frac{d_{ZA}}{1 + \sqrt{\frac{P_A}{P_Z}}} = \frac{10}{1 + \sqrt{\frac{20{,}000}{25{,}000}}} = 5.3$$

for center B,

$$d_{ZX} = \frac{6}{1 + \sqrt{\frac{10{,}000}{25{,}000}}} = 3.7$$

for center C,

$$d_{ZX} = \frac{16}{1 + \sqrt{\frac{16{,}000}{25{,}000}}} = 8.9$$

for center D,

$$d_{ZX} = \frac{20}{1 + \sqrt{\frac{4{,}000}{25{,}000}}} = 14.3$$

for center E,

$$d_{ZX} = \frac{8}{1 + \sqrt{\frac{4{,}000}{25{,}000}}} = 5.7$$

for center F,

$$d_{ZX} = \frac{9}{1 + \sqrt{\frac{15{,}000}{25{,}000}}} = 5.1$$

so that the market area for Z is given in Figure 5.2.

It is interesting to consider the relationship between β and the market area of place Z. Consider the market area boundary between Z and A that lies at 5.3 units from A when $\beta = 2.0$. As β gets larger and tends to infinity, $1/\beta \rightarrow 0$, so that

$$\left(\frac{P_A}{P_Z}\right)^{1/\beta} \rightarrow 1$$

and

$$d_{ZX} \rightarrow d_{ZA}/2$$

That is, as distance becomes a greater deterrent to interaction, the market area boundary moves outwards from place Z until the midpoint between the two centers is reached. The difference in populations between Z and A then plays no role in determining relative market areas—everybody shops at the nearest center and larger centers no longer appear more attractive for shopping.

Consider what happens to the same market area boundary when β gets smaller and tends toward zero. Then as $\beta \rightarrow 0$, $1/\beta \rightarrow \infty$ and

$$\left(\frac{P_A}{P_Z}\right)^{1/\beta} \rightarrow \left\{ \begin{array}{l} \infty \text{ if } P_A > P_Z \\ 0 \text{ if } P_Z > P_A \end{array} \right\}$$

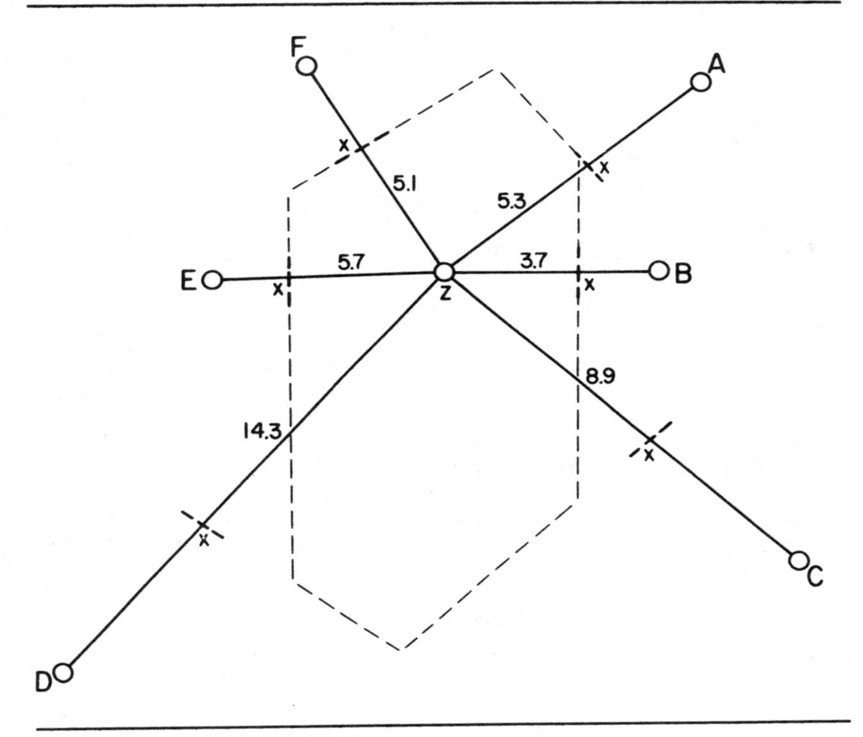

Figure 5.2 Market Area for Z

so that

$$d_{ZX} \rightarrow \left\{ \begin{matrix} 0 \text{ if } P_A > P_Z \\ d_{ZA} \text{ if } P_Z > P_A \end{matrix} \right\}$$

That is, when $\beta \rightarrow 0$, the deterrence of distance decreases until distance becomes inconsequential. At this point ($\beta = 0$), the market area boundary is determined solely by population size: People shop at the largest city so that if $P_A > P_Z$, A captures all of the market and $d_{ZX} = O$, whereas if $P_Z > P_A$, Z captures all of the market and $d_{ZX} = d_{ZA}$.

Example 3: Migration and Voting Patterns

Consider the very simple spatial system in Figure 5.3 (it may be a country with four states) in which zone centroids are denoted by a dot. Distances

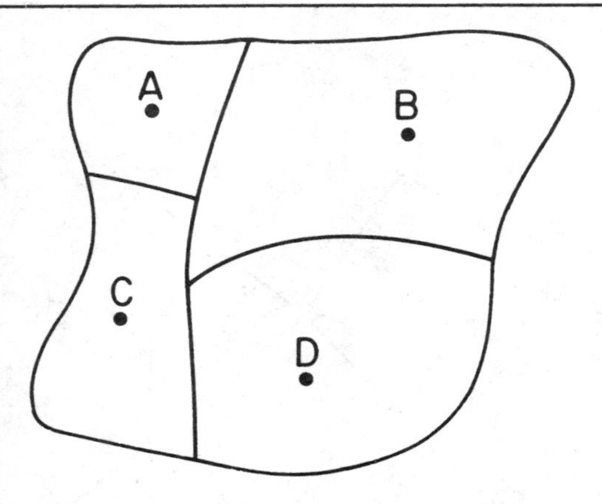

Figure 5.3 A Very Simple Spatial System

between these centroids can be thought of as measures of the average distance between zones. These distances are given below along with some other data on the four zones:

Distance matrix (miles);

	A	B	C	D
A	—	200	400	800
B		—	800	600
C			—	500
D				—

Zone	1982 Population	1982 Percentage of Democrats
A	100,000	45
B	400,000	75
C	50,000	45
D	200,000	55

Migration is taking place between the four zones and an analysis of previous migration flows (that is, the gravity model has been calibrated on existing flow data) shows that the model

$$\hat{T}_{ij} = (0.0003) \times (v_i^{1.0}) \times (w_j^{1.0}) \times (d_{ij}^{-1.0}) \qquad [5.3]$$

where v_i and w_j are defined in terms of population, gives reasonably accurate forecasts of the flows of migrants over a four-year period. It is also found that a person's decision to migrate is not related to his or her political beliefs so that the proportions of Democrats and Republicans in any flow are exactly the same as the proportions in the zone of origin of the flow.

Imagine that you are a political analyst and that you have been hired by the Democratic Party in zone C to forecast whether or not there will be a Democratic majority in that zone by 1986. We can now use the calibrated form of the gravity model given above and the data on each of the zones to produce a forecast of the percentage of Democrats in zone C in 1986.

Since the estimated values for the parameters and are equal in this instance, then $\hat{T}_{ij} = \hat{T}_{ji}$ for all pairs of i and j. That is, the predicted volume of migration from zone i to zone j will be the same as the predicted volume of migration from zone j to zone i. Note that such symmetrical predictions are a result of the estimates of α and λ (the parameters associated with v_i and w_j, respectively) being equal. If these estimates were unequal, then the predicted flows would not be symmetric unless the populations of i and j were equal. However, in this instance since $\hat{\alpha} = \hat{\lambda}$, we can write

$$\hat{T}_{AC} = \hat{T}_{CA} = \frac{.0003 \times 100{,}000 \times 50{,}000}{400} = 3{,}750$$

$$\hat{T}_{BC} = \hat{T}_{CB} = \frac{.0003 \times 400{,}000 \times 50{,}000}{800} = 7{,}500$$

$$\hat{T}_{DC} = \hat{T}_{CD} = \frac{.0003 \times 200{,}000 \times 50{,}000}{500} = 6{,}000$$

Consequently, the total emmigrants from C = total immigrants to C = 17,250. In 1982, zone C has a total of 50,000 × 45/100 = 22,500 Democrats, and a total of 50,000 × 55/100 = 27,500 Republicans.

The number of Democrats entering zone C from each of the other three zones is obtained by multiplying the number of migrants from each zone to C by the proportion of Democrats in the origin zone. A similar procedure can be followed to obtain the number of Republicans entering C from the other zones. These figures are as follows (after rounding):

	A	B	D	Total
Number Democrats entering C from:	1,688	5,625	3,300	10,613
Number of Republicans entering C from:	2,062	1,875	2,700	6,637

The total number of Democrats leaving C is given by multiplying the total number of migrants leaving C by the proportion of Democrats in C. A similar procedure is followed to derive the total number of Republicans leaving C. Thus, the total number of Democrats leaving C = 7,763, and the total number of Republicans leaving C = 9,487. Consequently, the net change in Democrats in zone C = number entering − number leaving = 10,613 − 7,763 = 2,850, and the net change in Republicans in zone C = 6,638 − 9,487 = −2,850. Therefore, in 1986 zone C will have 22,500 + 2,850 = 25,350 Democrats and 27,500 − 2,850 = 24,650 Republicans and the population of zone C will be composed of 50.7 percent Democrats and 49.3 percent Republicans.

Thus, you would forecast a slim Democratic majority for zone C in 1986. Notice, however, that we have made several assumptions in obtaining our forecast. One is that we have a "closed" system—that is, there is only movement within the system and there is no movement into or out of the system. People are not migrating into zone C from anywhere else but zones A, B, and D. Similarly, people in zone C are not migrating to any other places besides zones A, B, and D. In reality, this assumption is more tenable when the system under investigation is large. For example, if we were analyzing the effects of migration on voting patterns in the United States, then it may be reasonable to ignore immigrants from abroad (especially when we realize that immigrants are unlikely to have immediate voting privileges anyway).

Another assumption we made in our analysis is that the parameters of the model remain constant over time. Obviously, this assumption is more tenable over short than over long time periods. We have also assumed that our model is correctly specified and that we have calibrated it correctly. If the model is misspecified (that is, if the model contains the variables in an incorrect format or if the model does not contain all the relevant variables) or if we calibrate the model incorrectly, then our parameter estimates may be misleading and hence our forecasts are likely to be erroneous. Consider, for example, what happens if the actual distance-decay parameter were –2.0 and not –1.0 as we had estimated. Using the value of $\beta = -2.0$ in the model with α and λ both equal to 1.0 again and k = 0.1683 (k is adjusted so that the total outmigration from zone C remains the same) gives the following estimates:

$$\hat{T}_{AC} = \hat{T}_{CA} = 5{,}259$$

$$\hat{T}_{BC} = \hat{T}_{CB} = 5{,}259$$

$$\hat{T}_{DC} = \hat{T}_{CD} = 6{,}732$$

so that the total migrants from C = total migrants to C = 17,250. Consequently, the following numbers are predicted:

	A	B	D	Total
Number of Democrats entering C from:	2,367	3,944	3,703	10,014
Number of Republicans entering C from:	2,982	1,315	3,029	7,236

The total number of Democrats leaving C is 7,763, and the total number of Republicans leaving C is 9,487. As a result, the forecasted net change in Democrats is 2,251 and the forecasted net change in Republicans is −2,251, giving zone C 24,751 Democrats (or 49.5 percent of the population) and 25,249 Republicans (or 50.5 percent of the population) in 1986.

With the use of this different distance-decay parameter we would now forecast a slim Republican majority in zone C in 1986. This demonstrates the need for accurate model specification and accurate parameter estimation. The result also demonstrates the importance of distance-decay in explaining interaction patterns. The difference in the above two results is primarily caused by a decrease in the number of migrants predicted to move from zone B (a heavily Democratic zone) into C when $\beta = -2.0$ rather than -1.0. This decrease arises because the larger distance-decay parameter results from a greater perceived disutility of distance by migrants (the so-called "friction of distance" is greater). Consequently, there are fewer migrants who are likely to move between zones B and D, which are far apart, when the perceived disutility of distance is greater.

Example 4: Analyzing University Enrollments by State

Suppose a university wants to analyze its enrollment patterns by geographic area. It may want to do so, for example, in an attempt to identify regions in which it should recruit more heavily. Simply looking at the numbers of students enrolled in the university by state does not provide a complete picture. States with low populations and at long distances from the university can be expected to provide fewer students than states with large population and in close proximity to the university. However, it would be useful to know which states provide more (or fewer) students to the university than would be expected given their population and distance from the university. This can be done with the aid of a gravity model. In this instance, as we know the total enrollment at the university, an attraction-constrained gravity model is most appropriate. It is defined as

$$\hat{T}_{ij} = v_i^{\lambda} B_j D_j d_{ij}^{\beta} \quad [5.4]$$

where $\hat{T}_{ij}$ is the expected number of students attending univerity j from state i; v_i is a measure of the number of students available from state i; and d_{ij} is a measure of the separation of i and j (for example, the road mileage between the centroid of state i and the university at j). B_j is a balancing factor defined by

$$B_j = [\sum_i v_i^{\lambda} d_{ij}^{\beta}]^{-1} \quad [5.5]$$

which ensures that

$$\sum_i \hat{T}_{ij} = \sum_i \hat{T}_{ij} = D_j \quad [5.6]$$

In words, B_j ensures that the total expected enrollment at university j equals the actual total enrollment.

The difference between the actual and predicted enrollment $(T_{ij} - \hat{T}_{ij})$ is termed a *residual* and can provide information on student enrollment from particular states. Positive residuals indicate states providing more students to the university than would be expected given their population and distance from the university. Negative residuals indicate states from which fewer students attend the university than would be expected.

An example of this type of use of the gravity model is given in Figure 5.4, which depicts residuals from an analysis of graduate student enrollment at Indiana University. It appears that Indiana University receives more graduate students than would be expected from several states in the Northeast and from Kentucky, Kansas, Utah, and Washington, while it receives fewer graduate students than expected from Nevada, Texas, most of the South, Missouri, and West Virginia.

A similar analysis can be undertaken for any interaction pattern that converges on a single point. For example, since 1975 the newsletter of the Association of American Geographers has provided a breakdown of attendance by state at its annual conference. These data were used to calibrate an attraction-constrained gravity model for each of the seven locations of the annual conference between 1975 and 1981. The maximum-likelihood distance-decay parameters, along with R^2 values (giving an idea of how

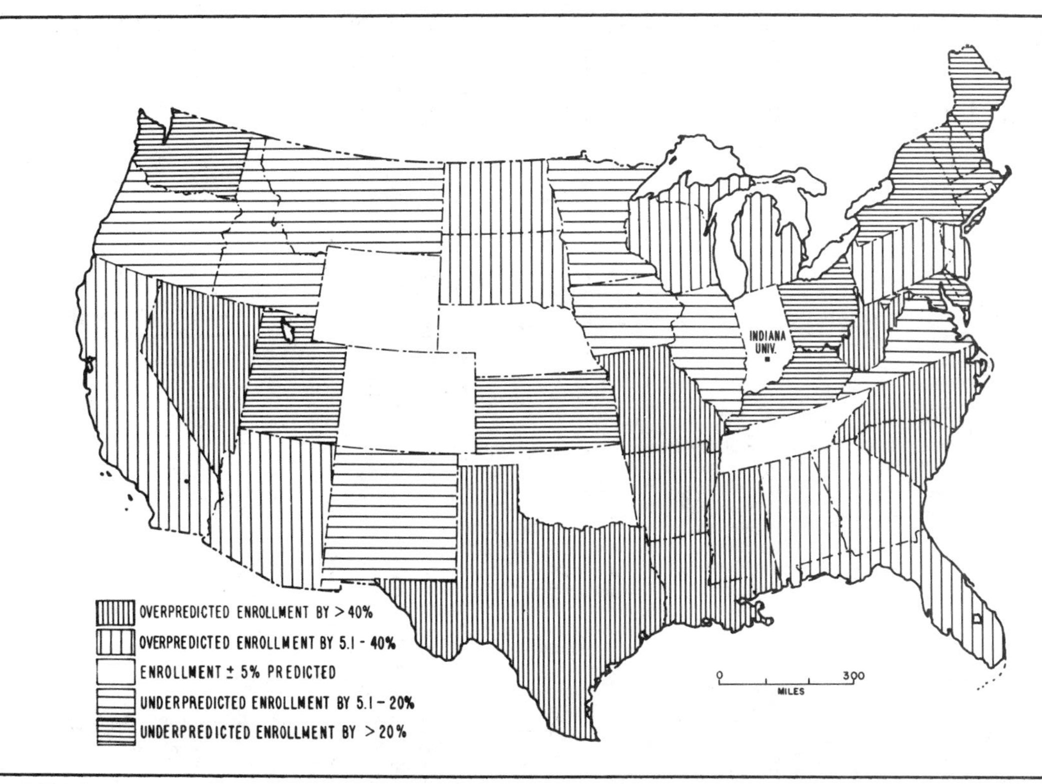

Figure 5.4 1980 Percentage Error in Model Predictions of Student Enrollment at Indiana University

well the calibrated model replicates the interaction patterns) are presented below. R^2 ranges from 0 indicating an extremely poor model fit to 1 indicating a perfect model fit.

Year	Host City of AAG Conference	$\hat{\beta}$	R^2
1975	Milwaukee	-.545	.822
1976	New York	-.367	.714
1977	Salt Lake City	-.460	.879
1978	New Orleans	-.483	.775
1979	Philadelphia	-.288	.888
1980	Louisville	-.527	.909
1981	Los Angeles	-.530	.984

The "population" of each state was taken as the state's membership of the Association of American Geographers. From an examination of the residuals for each of the meetings, the model consistently overpredicted attendance at the meetings from several states—Alabama, California, Maryland, Missouri, Tennessee, and Virginia. Similarly, attendance was consistently underpredicted from Colorado, Indiana, Iowa, Kansas, Michigan, Ohio, and Oklahoma. One conclusion from these results might be that geographers in the Midwestern states are more active in attending Annual Conferences than geographers from other states, especially those from the South-Central region of the country. Further questions then arise as to why such differences occur.

Example 5: Determining the Optimal Size of a Shopping Development

The following information is available on four neighboring towns, A, B, C, and D:

Town	Population	Department Store Floorspace (square feet)
A	100,000	25,000
B	10,000	0
C	50,000	0
D	5,000	0

The average weekly expenditure per person in department stores is estimated to be \$10.00 for each of the four towns. The location of the towns with respect to each other is given in Figure 5.5 where the numbers represent distances in miles.

A large department store (such as Penney's, Sears, etc.) is considering locating in town C where there is no department store at present but where the population could support one. The site for this development within town C has already been chosen and it is estimated that the average distance to this site from within the town is two miles. There are five sizes of department stores being considered: 1,000 square feet; 2,000 square feeet; 4,000 square feet; 5,000 square feet; and 10,000 square feet. You are brought in as a consultant to determine which of these sizes of store will maximize the firm's net pretax profits; these profits being defined by

$$\text{net pretax profits} = \text{total weekly sales} - \text{weekly costs} \qquad [5.7]$$

The weekly costs (salaries, purchase of goods, utilities, etc.) have been determined from existing stores and are derived from this formula:

$$\text{weekly costs (\$)} = 50 \times (\text{the size of the store in square feet})^{0.95}$$

which indicates that while total costs obviously increase as the size of the store increases, the costs/square foot decrease due to economies of scale.

One of the major reasons that you were brought in as a consultant was your experience with the gravity shopping model (a production-constrained gravity model):

$$\hat{c}_{ij} = \frac{E_i S_j^{\alpha} d_{ij}^{\beta}}{\sum_{j=1}^{n} S_j^{\alpha} d_{ij}^{\beta}} \qquad [5.8]$$

which is used to determine the level of sales in the proposed store in town C. In this model, $\hat{c}_{ij}$ is the predicted expenditure in department stores in j by people living in i; S_j is the department store floorspace in j; E_i is the total expenditure in department stores by all the inhabitants of i; and d_{ij} is the distance between i and j. From past experience with the model, you know that $\alpha = 1.0$ and $\beta = -2.0$ give accurate forecasts of the c_{ij}'s, the actual expenditures. The model can then be used to derive the figures in Table 5.1 from which it can be seen that the optimal size of the shopping

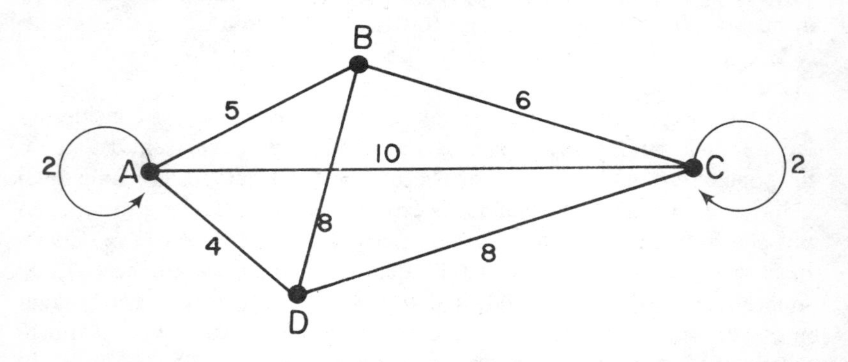

Figure 5.5 The Location of Four Towns

facility at C would be 4,000 square feet. Figure 5.6 represents the situation diagrammatically.

Both the weekly sales figures and the weekly costs increase with increasing facility size, but weekly sales level off much more quickly than do weekly costs. The following questions are left as an exercise for the reader:

(1) At what facility size would profit be exactly zero?
(2) Produce sales figures and percentages for $\beta = -1.0$ and $\beta = 0.0$. Explain the relationship between optimal facility size and β.
(3) For a given origin and when $\beta = 0.0$, why do the sales percentages remain constant as facility size increases?
(4) When $\beta < 0.0$, as the size of the proposed facility increases, the percentage of the trade derived from place C decreases and the percentages of trade from places A, B, and D increase. Explain this trend.
(5) It is a general result that regardless of the size of the facility, as β becomes less negative (i.e., tends toward zero), the percentages of total sales derived from places A, B, and D increase. Explain this trend.

Example 6: Locating a Facility for Maximum Patronage

This example demonstrates the use of a gravity-type model in determining the location of a facility that maximizes the patronage of the facility. The facility may be a public one such as a health care facility where the optimal location assures maximum patronage, or the facility may be a private one such as a fast-food restaurant where the optimal location assures maximum revenue. In the case of the fast-food restaurant, the location that maximizes revenue will also be the location at which profits are a maximum if land rents are constant. In the following example, we consider the case of locating an independent fast-food restaurant in a city that already has an established pattern of fast-food restaurants. In this type of problem, the

TABLE 5.1 Worksheet for Shopping Development Example

	Proposed Size of Shopping Development at C				
	1,000 sq. ft.	*2,000 sq. ft.*	*4,000 sq. ft.*	*5,000 sq. ft.*	*10,000 sq. ft.*
Weekly sales from C (in dollars)					
From A	1,597 (0.6)*	3,190 (0.9)	6,359 (1.5)	7,937 (1.8)	15,748 (3.2)
From B	2,703 (1.1)	5,263 (1.5)	10,000 (2.4)	12,194 (2.8)	21,737 (4.4)
From C	250,000 (98.1)	333,333 (97.2)	400,000 (95.6)	416,667 (94.9)	454,545 (91.5)
From D	495 (0.2)	980 (0.3)	1,923 (0.5)	2,380 (0.5)	4,547 (0.9)
Total	254,795	342,806	418,282	439,178	496,577
Weekly costs	35,397	68,383	132,107	163,302	315,479
Net pretax profits	219,398	274,423	286,175	275,876	181,098

*Figures in parenthesis are the percentages of total sales derived from each location.

predominant solution methods usually tend to involve some deterministic programming solution. That is, it is assured that people will travel to the nearest available restaurant. From all analyses of people's travel behavior this is an unrealistic assumption—people have a *tendency* to travel to the nearest facility but this is by no means the behavior pattern of everybody. There is usually a sizable proportion of people who, for a variety of reasons, do not patronize the nearest facility. To model such behavior, it is more accurate to employ a gravity-type model. Consider the city described in Figure 5.7, which is divided into 10 zones and contains 13 fast-food restaurants. The task is to decide where a new independent fast-food restaurant should be located so as to maximize its revenue. This can be done with the aid of a production-constrained competing destinations model (see Chapter 4),

$$S_{ij} = \frac{P_i E_i d_{ij}{}^{\beta} c_j{}^{\delta}}{\sum\limits_j d_{ij}{}^{\beta} c_j{}^{\delta}} \qquad [5.9]$$

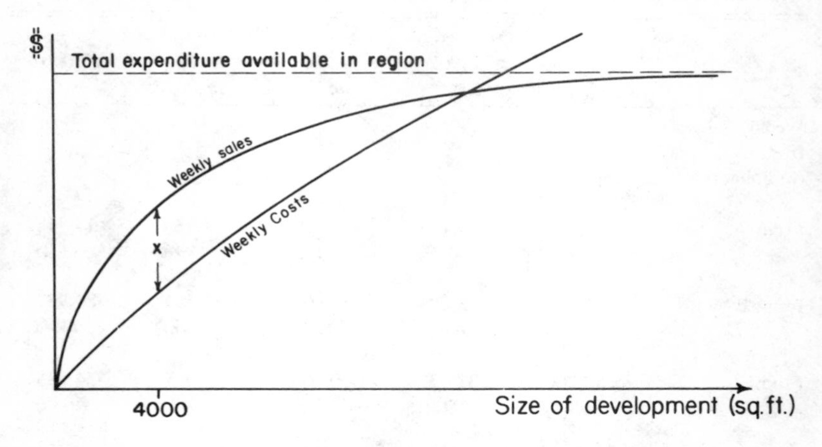

Figure 5.6 Optimal Size of Shopping Development

where S_{ij} is the expenditure on fast food at outlet j from zone i in a given time period: P_i is the population of zone i; E_i is the average expenditure per person on fast-food in zone i; d_{ij} is some measure of the separation between i and j; and c_j measures the accessibility of destination j to all other destinations. We assume that all of the destinations are equally attractive so that w_j can be omitted from the model and c_j can be defined as

$$c_j = q \cdot \sum_{\substack{k \\ k \neq j}} 1/d_{jk} \qquad [5.10]$$

where q is simply an arbitrary constant used to scale the accessibility values. Large values of c_j thus indicate facilities that are accessible to other facilities and that, consequently, face greater competition from other facilities.

Suppose that a household survey has been conducted in the city and people have been asked which fast-food restaurants they have patronized and how many times they have patronized each restaurant over a certain time period. They also have been asked how much they usually spend on fast-food over the same time period. These data, along with information on distances and

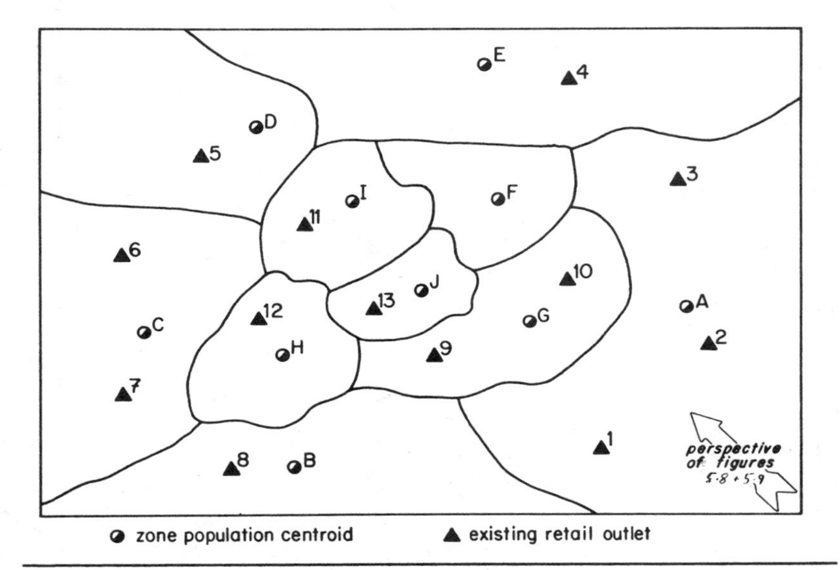

Figure 5.7 The Distribution of Population and Fast-food Outlets in a Hypothetical City

populations, allow the calibration of the interaction model. The procedure to find the optimal location for the new restaurant is then as follows:

(1) Place a grid over the city and read off the coordinates of each zone centroid and each existing restaurant.

(2) Select the location (0,0) as the initial guess at the optimal location. For this point, derive distances to all existing restaurants and to all zone centroids.

(3) Derive a set of accessibility measures and determine the values of $\hat{S}_{ij}*$ where j* represents the new facility. Determine the total revenue, T, at j* where

$$T_j* = \sum_i \hat{S}_{ij}* \quad [5.11]$$

(4) Repeat this process for all points on the grid and determine the location for which T_j* is at a maximum.

(5) Further information on the location of the new restaurant can be provided by mapping the revenue surface within the city. Such information can be valuable in determining alternative sites for the restaurant if the first choice site is unavailable for some reason (e.g., zoning restrictions, land unavailable, land values too high, etc.). Two such maps where β =-1.5, δ = 1.0 (agglomeration effect; see Figure 5.8) and where β = -1.5, δ = -1.0 (competition effects; see Figure 5.9) are included to demonstrate the influence of δ on the predicted revenue surface.

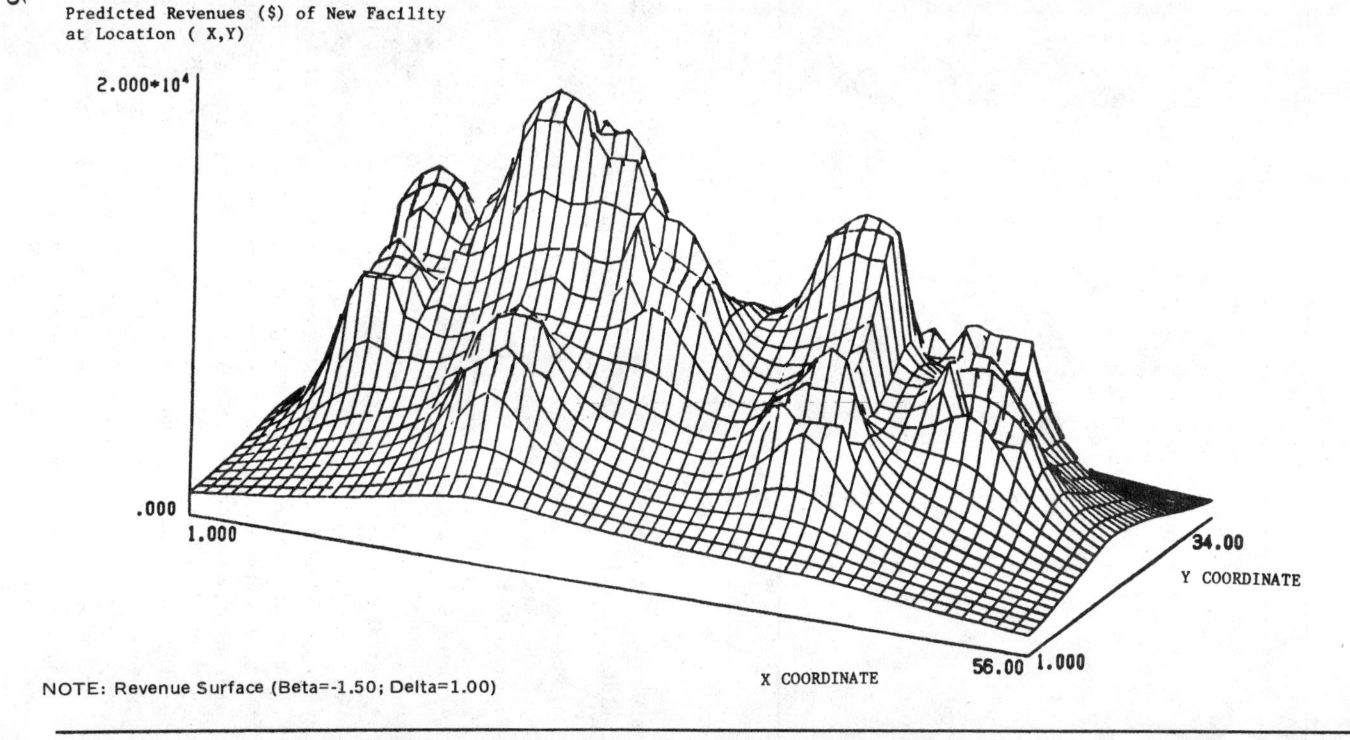

NOTE: Revenue Surface (Beta=-1.50; Delta=1.00)

Figure 5.8 Revenue Surface when Agglomeration Effects Exist Between Destinations

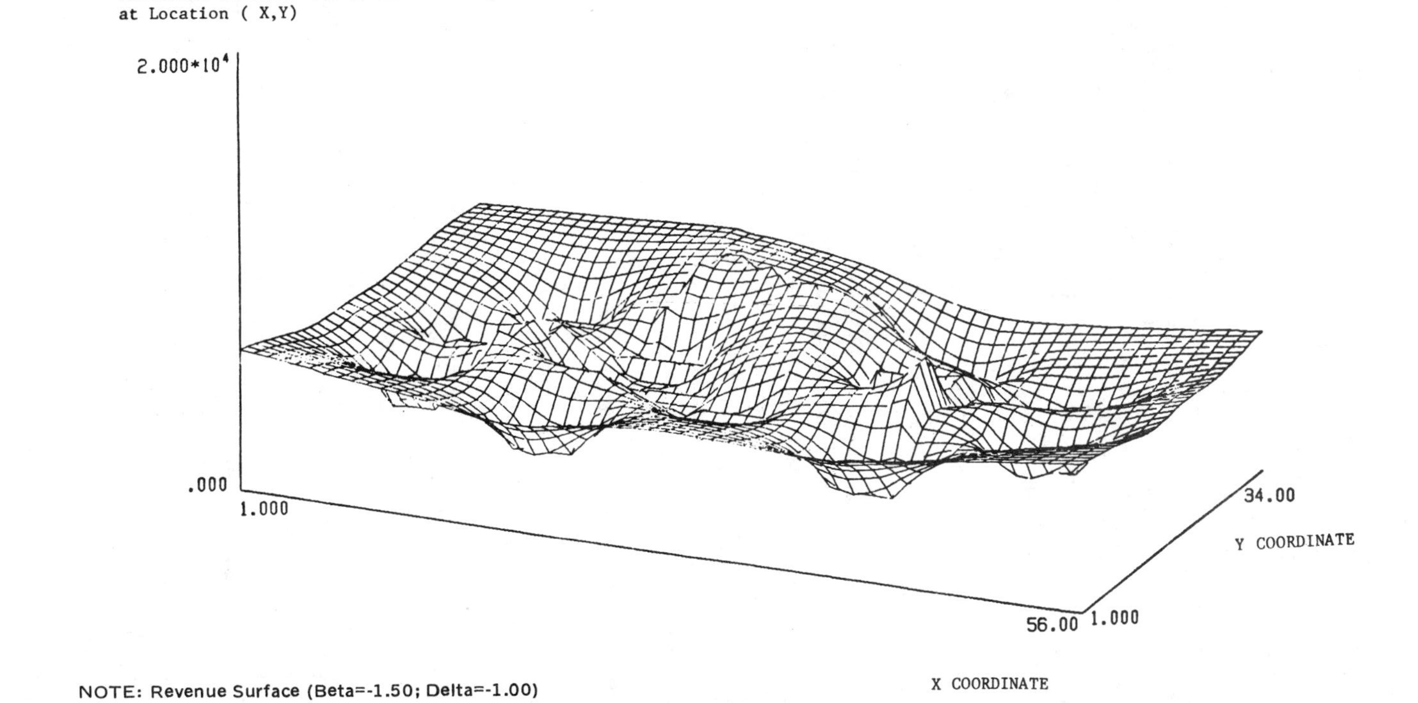

NOTE: Revenue Surface (Beta=-1.50; Delta=-1.00)

Figure 5.9 Revenue Surface when Competition Exists Between Destinations

NOTE

1. Although at the beginning of this example two situations were described in which this algorithm and the gravity model can be applied—the location of a fast-food restaurant and the location of a public health care facility—many other similar applications can be imagined. One other interesting application which, to our knowledge, has not been utilized yet, concerns the location of polling booths. Assuming that there is some distance-decay involved in voter turnout—that is, the nearer a person lives to a polling booth, the more likely is that person to vote—then the production-constrained gravity model can be applied as above. A political party could forecast the optimal location of a polling booth or booths in the sense of maximizing the proportion of people voting for that particular party!

6. OPERATIONAL CONSIDERATIONS

The use of any model is impacted by operational considerations and the gravity model is not an exception. Fundamental to all use considerations is data availability. Once data availability is established, issues of data format and quality must be addressed. Issues of prior research experience and the existence of standardized computer programs for running the model must then be considered. Finally, specialized situations must be examined for model modifications and special data requirements.

In 1974, Thompson reviewed the availability of spatial interaction data in the United States. He noted that the lack of good flow data was a handicap to spatial interaction research. A decade later the situation has not substantially improved. There is still a paucity of high quality data from the state and federal government and from private industry on the movement of people, information, and commodities. In spite of the increased sophistication of models and concepts that require more and more information, the data base for gravity modeling remains weak. Furthermore, even where such data do exist, their quality is often suspect.

In contrast to Thompson's assumptions, information on outflows from origins, inflows to destinations, and between cell interactions or dyads are not required of all gravity-type models. In fact the reason for developing the family of models in Chapter 2 was a recognition of the variability of data sets with regard to different origin and destination information. Indeed the purpose of the model is often to estimate the unknown information from the patterns of known information. However, to ensure that our estimates are correct or within reasonable levels of tolerance, we must have some fully developed interaction data set with which we can compare our results. Unfortunately, most pedagogical work on gravity modeling does not make reference to data needs or problems. We attempt, albeit briefly, to correct this situation.

It is not the purpose of this chapter to catalogue all federal, state, and local collections of interaction data. Although we will review some of the major U.S. government data sources, it should be remembered that this is only a small indication of the vast amount of data available. Besides data collected by other countries and sources of international flow data on airline passengers, immigrants, mail, and trade, a substantial amount of information is collected by private organizations and for proprietary purposes. To this must be added information from surveys by local government departments, volunteer organizations, and special interest groups. Although collected for special purposes, much of this information is often available if requested.

However, good data should have some basic characteristics that are of vital interest to users. A primary requirement of useful flow data is an accurate measurement of the magnitude of a flow for a specific period of time. Without data on magnitude over time the models cannot be calibrated with respect to rate and their outputs cannot be validated. Definitons and items of data need to be compatible and stable through time and space and also be sufficiently detailed so that interpretation is clear and consistent. Errors can often be reduced by careful sample design and by selecting large samples. Sampling error must be minimized because the multiplicative character of our models and the common use of exponents as power functions emphasize even relatively small errors.

Issues concerning the scale of areal units and the heterogeneity of place characteristics, issues concerning administrative bias at various levels of government, as well as issues concerning the efficient design of spatial units must be recognized and examined before data are utilized (Berry 1968). The appropriateness of various areal units (census tracts, counties, states) in terms of scale and function is not unique to spatial interaction data—it is an ongoing geographic problem. Spatial interaction models are affected by alternative zoning of origin and destination areas, and different aggregation rules can produce significantly different results. Openshaw (1977) demonstrated the impact of alternative zoning on spatial interaction models, while Haynes and Storbeck (1978) examined the statistical basis for some of Openshaw's results by using an information measure of uncertainty. Clearly, a gravity modeler wishing to optimize model performance must be sensitive to the areal base upon which aggregate origins and destinations are based. Often, available data do not allow much flexibility in this regard. Usually the only option is to alternate zone aggregation patterns and examine model performance in light of the fact that data on the character of individual zones or areal units are often limited.

Much of the information gathered for spatial interaction analysis at the national level is gathered by the federal government. The U.S. federal government spends over $750 million per year on data collection and preparation by its principal statistical and line agencies. This magnitude of involvement has not produced the kind of stability that most users would expect or desire. There are substantial agency-to-agency variations in definitions, sampling, and measures of reliability. There are a host of private data collection services used by government agencies to supplement their own and other government data collection activities. The Presidential Commission on Federal Statistics (1971) commented on this lack of coordination in data-gathering activity and statistical policy, recommending a substantial overhaul of the system. The disastrous implications of this poor statistical situation for the development of forecasting models and policymaking are described fully in House et al. (1976) and House and Williams (1978).

Next we will examine data issues associated with commodity flows, interurban passenger movements, intraurban transportation, and migration.

Commodity Flows

Although geographers, in particular, have a good record in the analysis of U.S. commodity flows, there has still been limited use of available data. Up to this point the focus has been on aggregate relationships, such as total flows by weight and value, rather than on commodity type, such as fabricated steel or glass. (Ullman 1957; Perle 1964; Black 1973). Economists' interests have been limited to describing (Spiegelglas 1960), forecasting (Meyer and Straszheim 1971), or providing the basis for interpreting other models (Moses 1955; Pfister 1961; Olson 1972). Overall, the use of gravity models for U.S. commodity flow analysis has been substantially limited (Black 1971; 1972; Knudsen 1983). However, Berry (1966) has modeled commodity flows in the Indian economy; Chisholm and O'Sullivan (1973), Pittfield (1978), and Gordon (1978) have modeled freight flows in the economy of the United Kingdom; and Linneman (1966) has utilized origin-specific gravity formulations for examination of international trade.

The sources of interstate freight flows in the United States vary by mode of carrier although the most widely used system is the sample of Waybills collected and reported by the U.S. Interstate Commerce Commission, which is particularly valuable for rail and trucking systems. Due to the size of the sample, regional and interregional gross flow information ranges from good to adequate, but flows for highly disaggregated dyads or those disaggregated by commodity type are extremely small and hence are subject to substantial sampling variability. Rosander (1965) provides one of the few evaluations of the quality of the annual Interstate Commerce Commission's Waybill sampling information upon which many studies have been based.

Data on air cargo movements are even more limited but for less understandable reasons considering that most flights in the United States cross interstate lines. This problem is not likely to be remedied in the future given the demise of the Civil Aeronautics Board which collected some limited but important information in this area. However, most of this freight information was not of the city pairs or origin-destination variety. A noteworthy exception was the frequent flow information reported by commuter airlines. Highway freight transportation and air freight now account for 30 percent of total U.S. freight movements with important specialized characteristics. To have poor flow coverage of these vital movements is of great concern to many transportation analysts. Hence the prospect of declining rather than increasing information on these flows is disturbing.

Although there have been some state and metropolitan studies of local freight movements (e.g., Chicago Transportation Study 1965), they are rare, and analysis of the work in the United States with respect to the above has been limited to the transportation planning groups at the University of Pennsylvania (Allen et al. 1983; Friesz et al. 1981; Friesz and Vinton 1982; Harker 1983) and at the University of Illinois (Southworth 1982; Southworth et al. 1983a, 1983b).

Other commodity-related statistics are collected by the U.S. Army Corps of Engineers and the Maritime Administration, particularly as they reflect water borne freight (Patton 1956). The corps' information is published annually by transportation segment and some origin-destination data are available for shipping and receiving regions for specific ports and waterways for at least six major commodity groups. The national Maritime Administration data are for 10 large regional units by commodity; for aggregate information, a port-to-port matrix can be generated.

Specific criticisms of commodity information have been leveled by the U.S. General Accounting Office (1974) and continue to be relevant. These include the following:

(1) Data bases of relevant information needed for statistically reliable forecasts have been neglected.

(2) Production, consumption, and price information for key industries and mineral products are unavailable to the government, except through industry intermediaries.
(3) Modern statistical methods and research techniques have not been used to make commodity forecasts.

Hence it can be seen that the existence of commodity forecasts does not imply quality in terms of data reliability or in some cases even validity. The implication of this is clear. Researchers use commodity data at their own risk and need at the very least to go back to primary sources and do substantial reliability and validity testing.

Interurban Passenger Movements

The primary modes for interurban passenger movement in the United States are air, bus, and rail. Water passenger transportation is insignificant except in Seattle and New York, and even there it borders on trivial. Interstate passenger rail movements are small relative to other passenger flows, and no national flow data are available. Individual railroads have traffic information, and metropolitan transit authorities collect information in some areas (e.g., New York and Chicago). Amtrak has collected some rail passenger information; and as Amtrak passengers constitute a growing proportion of all rail passengers (growing in a relative, not absolute, sense), this may eventually be a useful data set. Bus transportation of passengers is important locally, regionally, and in terms of interstate movements. However, spatial flow data are almost nonexistent. Some metropolitan communities manage to collect such information for publicly owned carriers, but not even gross total flows are well organized and useful. Most gross total flow information is by segment of trip with passenger or trip specific origin and destination information almost unknown. The U.S. Transportation Census is not helpful in terms of interurban origin and destination flows for passengers.

Air passenger transportation is, in relative terms, a better situation. The annual survey of air passengers' ticketed on certified carriers provide reliable and accurate flow volumes from a 10 percent sample. There are data limitations concerning passenger characteristics, such as income as well as limited information on trip purpose, and single- as opposed to multistop travel. However, these are among the best U.S. passenger transportation flow data available. As a consequence, there is substantial experience in modeling this information. In terms of aggregate assessment of interaction, the work of Ikle (1954) and Hammer and Ikle (1957) should be mentioned; interurban analysis of the impact of city size can be found in Taaffe (1956), Richmond (1957), Lansing (1961), and Long (1968, 1969). Taaffe (1959) and Howrey (1969) used these data for forecasting purposes. Other forecasting applications that focused on regression and time-series methodology include Richmond (1955) and Brown and Watkins (1968). Direct use of this data primarily for gravity modeling can be found in Alcaly (1967), Long (1970), and Fotheringham (1981, 1983a). However, an important final note is that the source of this information, the Civil Aeronautics Board, is disappearing as a government agency as part of the U.S. government's deregulation program. Hence, the source of this information may disappear along with the regulated environment that provided the basis of these studies.

Intraurban Transportation

The intraurban transportation category of spatial interaction is the most widely modeled by gravity formulations (Quandt 1965). As early as 1965, the U.S. Department of Commerce published directions on using gravity models for different size urban areas (U.S. Department of Commerce, 1965); the U.S. Department of Transportation (DOT) issued a report on the calibrating and testing of these models (1973). The Urban Mass Transportation Administration included elements of the gravity model in its standardized planning system (DOT 1972) which in turn became an integral part of DOT's PLANPAC and BACKPAC computer programs used for transportation planning. However, the flexibility of the model and the desire of DOT's planners to use the output of the model to be able to evaluate alternative transportation needs led to the standardization of model parameters by city size and to the maintenance of the computer program in machine language so that it cannot easily be adapted or adjusted. An excellent overview of the attempts to increase the transferability of some parameters of the model is presented in the *Quick-Response User's Guide* (Sosslau et al. 1978). In DOT's 1981 version of the Urban Transportation Planning System the PLANPAC/BACKPAC transport planning process was replaced by the less costly, but equally rigid, Quick-Response System program.

DOT's Federal Highway Administration has generated substantial data on intraurban transportation movements, both directly through its metropolitan origin and destination surveys (1950s through 1970s) and indirectly by funding collection activities of research organizations and state and local governments. Problems in the use of these intraurban surveys and statistical information on data collection, specialized sampling, and error problems are extensively reviewed in DOT's many reports. Conceptual issues, however, are not widely discussed. For example, there is usually no attempt to collect information on walking trips. This means, in general, short distance travel is substantially underestimated and single purpose short distance trip information is likely to be invalid. Similarly it was not until late in the survey series that household income information was collected.

The journey-to-work travel pattern is a major focus of intraurban transportation studies, but in the United States most of this work is based on either DOT's origin-destination surveys or local transportation authority surveys, an example of the latter being the Chicago Area Transportation Study. This information has not been widely used in gravity-related frameworks for modeling urban travel (Getis 1969; Wheeler 1972) but similar data have been used elsewhere, for example Canada (Hutchinson and Smith 1979), London, and Denmark (Clark and Peters 1965).

When data are missing either in terms of origins, destinations, or between-cell interactions, estimation methods can be utilized. However, estimation is never a complete substitute for survey information. On the other hand, survey resource requirements are large and usually only available intermittently. Further, although private survey collections continue there is little to encourage the hope that a national intraurban transportation survey system will be likely to start or continue for long. Hence these short-term and often private or local surveys, ranging from traffic counts to windshield surveys by municipal government, will continue to be important.

Migration

The detailed review of migration data by Isserman, Plane, and McMillian (1982) is the most up-to-date and comprehensive evaluation of federal sources presently available. The five sources for U.S. migration data are the U.S. Census of Population, the Current Survey of Population, (CSOP), the Annual Population Estimate Program (APEP), the Internal Revenue Service's tax records (IRS), and the Social Security Administration's records (SSA). Although Isserman et al. reviewed the adequacy of these data sets for length of migration period, timeliness, and forecasting, we are primarily interested in flow coverage and geographic detail for our gravity modeling purposes.

The U.S. census collects gross flow information from a sample survey that has declined in size from 100 percent in 1940 to 10 percent in 1980.

Geographic information is disaggregated to the county level and can be grouped to standard economic areas (SEAs) and state levels with good demographic detail on age, sex, race, income, and occupation. A major problem with this source of migration data is the use of different time spans (usually five years but one year in 1950) producing comparability problems due to return and repeat migration (Keyfitz and Philipou 1981).

The Annual Population Estimate Program of the census gives only net totals for migration disaggregated to the state and county level with poor demographic detail. It uses a residuals-derivation method: net migration = estimated population − census year population − births + deaths. This residuals-derivation method uses three separate population estimates that individually have low levels of error; by averaging, this error can be kept within acceptable bounds for the total population. However, the entire error in the estimate is forced into the smaller net migration estimate by the residuals equation. Furthermore, population estimate errors tend to be larger for rapidly changing areas than for stable areas. Finally, this system is not intended to be used as a migration time series, and unpublished "after-the-fact" adjustments are made decennially and at arbitrary intermediate periods.

The Current Survey of Population began in 1948 and is part of the U.S. Bureau of the Census P-20 series. It is a small sample survey of between 45,000 and 70,000 U.S. residents and provides overall movement rates and gross flows with demographic, occupational, and relative income information by census regions only. Due to both the size of the sample and the lack of detailed geographic coverage it is of little value for gravity modeling purposes.

The final two federal migration information sources, the IRS and SSA, using matched administrative records, suffer from sample problems and lack of complete coverage. IRS provides gross flow information with variable migration periods (one, two, three and five years). It is a 100 percent sample of all tax filers, which does not provide complete population or even wage earner coverage. However, the population coverage is better than one at first suspects. For 1980, including filers and claimed dependent exemptions, about 94 percent of the U.S. population was covered. No demographic information is available, but geographic detail to the state level is possible. The IRS series is discussed in some detail by Engels and Healy (1981). Cartwright (1978; Cartwright and Armknecht 1980) has reviewed the SSA migration data, and the U.S. Bureau of Economic Analysis (BEA) has provided an earlier review (1976). The SSA's program follows individual records through time. It began as a 1 percent sample and is now a 10 percent sample. This means that gross flows at aggregated regions (states, counties, and BEA regions) with detailed demographic information is available but only for appropriately "covered" workers (about 90 percent of all U.S. wage earners).

In all of the above cases, where a sample is modest to small in size (10 percent) and good geographic detail is available, with spatial disaggregation the flow magnitudes become very small; and although overall accuracy may be good, individual dyad estimates can vary enormously. A review and assessment of particularly the temporal component or time series elements of the matrices for the IRS and CSOP data series is provided by Rogerson and Plane (in press).

There has been a long history of the use of gravity models in the analysis of U.S. migration data. Greenwood and Sweetland (1972) review and analyze the determinants of migration in terms of metropolitan flows, while Plane (1981) and Tobler (1983) have each utilized a new and different form of the gravity model in assessing migration. Goodchild and Smith (1980) and Smith and Clayton (1978) examine U.S. migration streams in terms of particular gravity model characteristics, while Smith and Slater (1981) try to incorporate concepts of information flows into gravity models of U.S. migration patterns. Clark and Ballard (1980) specifically examine the role of origin and destination characteristics on out-migration flows. The literature in terms of gravity model applications to migration is vast and is only cited in passing here.

Special Applications

There are three special applications of gravity models that should be mentioned in terms of operations considerations. These are the modeling of recreation trips, the modeling of information flows, and the modeling of social-interaction travel. Each of these use special data and are applications of the general models to problems that are not treated as examples in Chapter 4.

Recreation trips can be long distance and low frequency for such things as vacation travel, and hence learning characteristics and repeat patterns may be quite different than gravity models of journey-to-work or business travel. Furthermore there are a variety of models of recreation trip distributions as outlined by Baxter and Ewing (1981). Problems associated with modeling this travel behavior in a gravity format are reviewed by Ewing (1982) using Scottish data. Specific issues in calibrating recreation trip data have been outlined by Goodchild (1975), and for the production-constrained model in particular they have been discussed by Baxter and Ewing (1979).

In a different context, Goodchild and Booth (1980) use an interaction model of short distance, high frequency intraurban recreation behavior in order to locate optimally a new public recreation facility. The data used for recreation-trip modeling are based either on surveys of people arriving at a destination or of intentions of going to a particular destination from a survey at the place of trip origin. The former information base is the most common; many parks and recreation centers gather this kind of data. Basically these are user-oriented models and as such are of immense value to the private or public agency that offers the destination service.

Information flows and economic development have been related to each other using gravity formulations (Tornquist 1968). Often this took the form of measuring the potential for communication by examining transportation access or complexity. Such aggregate relationships have been demonstrated in developing country contexts in Africa (Riddell 1968) and in Southeast Asia (Leinbach 1973). In Latin America, Gauthier (1968) has demonstrated that if transportation is a surrogate for information flows then economic growth is closely related to this pattern. In a parallel (but developed-country) context, Nelson (1959) and Haynes and Ip (1971) demonstrated that variations in a transportation measure of communication complexity were closely related to the scale and distance parameters of the gravity model. One of the clearest pedagogical developments of spatial information as a learning process was enunciated by Gould (1975). In Gould's work the gravity model parameters of scale and distance are carefully explored. Finally it might be noted that the Bell Telephone System in the United States and Canada have been major users of gravity models for assessing and projecting long distance communications traffic. It would appear that the use of gravity modeling in the spatial management of information flows is likely to continue to increase rapidly.

Social travel behavior has been less widely studied than simple unitary purpose trip behavior. In some respects, this is due to its multiobjective and multiattribute character. The work of Wheeler and Stutz (1971) and Stutz (1976) demonstrate the variety of concerns that exist under this important but loose label. Stutz (1973) reports that the effect of the "distance decay" or "friction of distance" parameter on trips relates more to social choice contacts than to visits to relatives. Social choice contacts including travel to friends and neighbors have steep distance-decay curves, while trips to relatives, which are less a function of social choice, have a relatively flat distance-decay function. This reinforces the early findings of Hägerstrand (1966) and Morrill and Pitts (1967) that distance between marriage partners—outcomes of social contacts—has an important and significant distance-decay parameter: the probability of marriage being closely associated with propinquity. One of the major difficulties in the social trip aspect of interaction research is the need to specify the correct scale or attraction parameters. These obviously vary substantially with trip purpose.

When home is the locus around which the interaction takes place, the distance decay results are likely to be quite different than if it were an away from home job site. This may be an important component in some aspects of traditional sex differences in social travel interaction. This area of social travel interaction is a difficult but fruitful field for expansion of gravity-model-based research

Planned Economies

The use of gravity models has been dominated by residential location or retail/service system considerations in free market economies. However, Korcelli (1976) points out that planned economies have also been utilizing these approaches. Gravity models in planned economies are obviously also used for service distribution considerations and population access assessment. (Domanski and Wierzbicki 1983). However, the most intriguing emphasis has been on journey-to-work modeling as outlined by Golc (1972). A clear incorporation of a potential model of attraction is found in Chauke (1960) and Shershevsky and Shnurov (1965). Furthermore the transfer of many of these ideas to East European planners, with particular reference to the Lowry model, was stimulated by the American-Yugoslav Project in Ljublana 1968-1976 (Zipser 1973). Spatial interaction models dealing with the exchange of materials and information in the planned economy have received some attention as it affects linkages between old and new industrial production complexes (Bandman 1973; Zagozdzon 1973, 1976). An extensive review of Soviet settlement planning with some references to spatial interaction concepts is given by Sokolow (1975). One of the more active research frameworks for the use of these models appears to be in Poland (Chojnicki 1966; Korcelli 1975; Dobrowolska 1976; Polarczyk 1976; Domanski 1979).

Theory and Application

Operational considerations affect the use of the gravity model in a number of ways. Previous experience provides some evidence of their usefulness and limitations. Data availability helps determine which of the family of gravity models discussed in Chapter 2 is most relevant. Data organization and quality provides information on the levels of disaggregation that are possible at different levels of confidence. Next, special applications provide some evidence of expanding areas of gravity model uses and some of the special requirements of these areas. Finally the use of the gravity model in planned economies gives some evidence of the breadth of the appeal of these models.

From the data perspective, our initial impression of a large variety of high quality data, widely disseminated and available, appears on closer examination, to be incorrect. Sample size is usually too small; and sample error, although small in the aggregate, is often large and unpredictable when cell-to-cell interactions are examined. The data availability for widespread validation and reliability testing of the model is poor and not likely to improve.

Gravity models have evolved through interactions with real-world considerations of data availability and predictive accuracy. They have also evolved through theoretical development and purposeful interpretation. These two kinds of research forces are not independent of each other. We moved from a deterministic social physics framework to a probabilistic individual decision theory framework by application considerations (Senior 1979). Finally, we moved to the basic top-down statistical estimation framework of the entropy model by another series of application considerations. None of these movements would have been possible if the theoretical frameworks did not exist beforehand, but none of these movements would have been required if the applications had not necessitated the change. Theory and application, therefore, are important interactive elements that are fundamental operational considerations in the continued evolution of any modeling methodology and are particularly important with respect to the gravity model.

REFERENCES

Alcaly, R. E. 1967. Aggregation and gravity models: some empirical evidence. *Journal of Regional Science* 7: 61-73.

Allen, B., et al. 1983 The experience with trucking deregulations: the New Jersey case. Conference on Regulatory Reform in Surface Transportation, U.S. D.O.T. Conference, Syracuse University, New York.

Bandman, M. 1973. Optimizatsya prostranstrennoy struktury khozyaystra ekonomitheskogo rayona. In *Primeyenye Modeley dlya Razrabotki Skhemformirovanya TPK*, pp. 149-164. Novosibirsk.

Batty, M. 1978. Reilly's challenge: new laws of retail gravitation which define systems of central places. *Environment and Planning A* 10: 185-219.

Baxter, M. J., and Ewing, G. O. 1979. Calibration of production-constrained trip distribution models and the effect of intervening opportunities. *Journal of Regional Science* 19: 319-330.

———. 1981. Models of recreational trip distribution. *Regional Studies* 15: 327-344.

Berry, B. J. L. 1966. *Essays on commodity flows and the spatial structure of the Indian economy.* Research Paper, No. 111. Chicago, University of Chicago, Department of Geography.

———. 1968. *Metropolitan area definition: a re-evaluation of concepts and statistical practice.* Working Paper No. 28;. Washington, DC: U.S. Bureau of the Census.

Black, W. R. 1971. The utility of the gravity model and estimates of its parameters in commodity flow studies. *Proceedings of the Association of American Geographers* 3: 28-31.

———. 1972. Interregional commodity flows: some experiments with the gravity model. *Journal of Regional Science* 12:107-118.

———. 1973. Toward a factorial ecology of flows. *Economic Geography* 49: 59-67.

Brown, S. L., and Watkins, W. S. 1968. The demand for air travel: a regression study of time-series and cross-sectional data in the U.S. domestic market. *Highway Research Record* 213: 21-34.

Carey, H. C. 1858. *Principles of Social Science*, vol. 1, Philadelphia: Lippincott.

Cartwright, D. W. 1978. Major geographic limitations for CWHS files and prospects for improvement, *Review of Public Data Use* 7: 16-26.

Cartwright, D. W., and Armknecht, P. A. 1980. Statistical uses of administrative records. *Review of Public Data Use* 8: 13-25.

Charnes, A., Raike, W. M., and Bettinger, C. O. 1972. An extremal and information-theoretic characterizations of some interzonal transfer models. *Socio-Economic Planning Sciences* 6: 531-537.

Chauke, M. O. 1960. K voprosu isucheniya zakonomernosti vnutrigorodskogo i prigorodskogo rasseleniya. *Problemy Sovetskogo Gradostreitelstra* 8.

Chicago Area Transportation Study. 1959-1968. *Survey findings.* Chicago: Chicago Area Transportation Study.

Chisholm, M., and O'Sullivan, P. 1973. *Freight Flows and Spatial Aspects of the British Economy.* Cambridge: Cambridge University Press.

Chojnicki, Z. 1966. Zastosowanie modeli grawitacji i potencjalu prgestrzenno-ekonomicznych. *Studia KPZK PAN 15,* Warsaw.

Clark, C., and Peters, G. H. 1965. The intervening opportunities method of traffic analysis. *Traffic Quarterly* 19: 104-115.

Clark, G., and Ballard, K. 1980. Modeling outmigration from depressed regions: the significance of origin and destination characteristics. *Environment and Planning A* 12: 799-812.

Clark, J. R. 1979. Measuring the flow of goods with archaeological data. *Economic Geography* 55: 1-17.

Cliff, A. D., Martin, R. L., and Ord, J. K. 1975. Map pattern and friction of distance parameters: reply to comments by R. J. Johnston;, et al. *Regional Studies* 9: 285-288.

Converse, P. D. 1949. New laws of retail gravitation. *Journal of Marketing* 14: 379-390.

Curry, L. 1972. A spatial analysis of gravity flows. *Regional Studies* 6: 131-147.

Curry, L., Griffith, O., and Sheppard, E. S. 1975. Those gravity parameters again. Regional Studies 9: 289-296.

Dobrowolska, M. 1976. The growth pole concept and the socio-economic development of regions undergoing industrialization. *Geographia Polonica* 33: 83-101.

Dodd, S. C. 1950. The interactance hypothesis: a model fitting physical masses and human groups. *American Sociological Review* 15: 245-257.

Domanski, R. 1979. Accessibility, efficiency and spatial organization. *Environment and Planning A* 11: 1189-1206.

Domanski, R., and Wierzbicki, A. P. 1983. Self-organization in dynamic settlement sytems. *Papers of the Regional Science Association, Pacific Conference* 51: 141-160.

Engels, R. and Healy, M. 1981. Measuring interstate migration flows: an origin-destination network based on internal revenue service records. *Environment and Planning A*, 13: 1345-1360.

Ewing, G. O. 1982. Modeling recreation trip patterns: evidence and problems. *Ontario Geography* 19: 29-56.

Foot, D. 1981. *Operational Urban Models: An Introduction.* New York: Methuen.

Fotheringham, A. S. 1981. Spatial structure and distance—decay parameters. *Annals of the Association of American Geographers* 71: 425-436.

———. 1983a. A new set of spatial interaction models: the theory of competing destinations. *Environment and Planning A* 15: 15-36.

———. 1983b. Some theoretical aspects of destination choice and their relevance to production-constraining gravity models. *Environment and Planning A* 15: 1121-1132.

———. 1984a. Spatial flows and spatial patterns. *Environment and Planning A* 16: 529-543.

———. 1984b. Spatial competition and agglomeration in urban modeling. *Environment and Planning A*, in press.

——— and Dignan, T. 1984. Further contributions to a general theory of movement. *Annals of the Association of the American Geographers*, in press.

Fotheringham, A. S., and Webber, M. J. 1980. Spatial structure and the parameters of spatial interaction models. *Geographical Analysis* 12: 33-46.

Friesz, T. L., Gottfried, J., Brooks, R. E., Albin, J. Z., Tobin, R., and Melesk, S. A. 1981. *The northeast regional environmental impact study: theory validation and application of a freight network equilibrium model.* Argonne National Laboratory Report ES-120. Argonne, IL: Argonne National Laboratory.

Friesz, T. L., and Vinton, P. A. 1982. *Economic and computational aspects of freight network equilibrium models, a synthesis.* Regional Science Working Paper No. 70. Philadelphia: University of Pennsylvania.

Gauthier, H. 1968. Transportation and the growth of the Sao Paulo economy *Journal of Regional Science* 8: 77-95.

Getis, A. 1969. Residential location and the journey from work. *Proceedings of the Association of American Geographers* 1: 55-59.

Golc, G. A. 1972. Vliyanye transporta na prostranstvennoe razvitye goorodov i aglomeratsii. *Problemy Sovremennoy Urbanizatsii,* pp. 159-190. Moskva: Statiska.

Goldner, W. 1971. The Lowry model heritage. *Journal of the American Institute of Planners* 37: 100-110.

Goodchild, M. F. 1975. *Fitting the general spatial interaction model to recreation trip data.* CORD Technical Note. No. 35. Ottawa: Parks Canada.

Goodchild, M. F., and Booth, P. J. 1980. Location and allocation of recreation facilities: public swimming pools in London, Ontario. *Ontaria Geography* 15: 35-51.

Goodchild, M. F., and Smith, T. R. 1980. Intransitivity, the spatial interaction model and U.S. migration streams. *Environment and Planning A* 12: 1131-1144.

Gordon, I. R. 1978. Distance deterence and commodity values. *Environment and Planning A* 10: 889-900.

Gould, P., 1975. Acquiring spatial information. *Economic Geography* 51: 87-99.

———. and White, G. 1974. *Mental maps.* London: Penguin.

Greenwood, M. J., and Sweetland, D. 1972. The determinants of migration between standard metropolitan statistical areas. *Demography* 9: 665-681.

Hägerstrand, T. 1966. Aspects of the spatial structure of social communication and the diffusion of information. *Papers of the Regional Science Association* 16: 27-42.

Hallam, B. R., Warren, S. E., and Renfrew, C. 1976. Obsidian in the Western Mediterranean: characteristics by neutron activation analysis and optical emission spectroscopy. *Proceedings of the Prehistoric Society* 42: 85-110.

Hammer, C., and Ikle, F. C. 1957. Intercity telephone and airline traffic related to distance and the propensity to interact. *Sociometry* 10: 306-316.

Hansen, W. G. 1959. How accessibility shapes land use. *Journal of the American Institute of Planners* 15: 73-76.

Harker, P. T. 1983. *Prediction of intercity freight flows: theory and applications of a generalized spatial price equilibrium model.* Ph.D. dissertation, Department of Civil Engineering, University of Pennsylvania.

Haynes, K. E. 1968. *Movement-behavior theory: an overview and summary.* Pittsburgh: Consad Research Corp.

Haynes, K. E. and Phillips, F. Y. 1982. Constrained minimum discrimination information: a unifying tool for modeling spatial and individual choice behavior. *Environment and Planning A* 14: 1341-1354.

Haynes, K. E., Phillips, F. Y., and Mohrfeld, J. W. 1980. The entropies: some roots of ambiguity. *Socio-Economic Planning Sciences* 14: 137-145.

Haynes, K. E., Poston, P. L., and Schnirring, P. 1973. Intermetropolitan migration in high and low opportunity areas: indirect tests of the distance and intervening opportunities hypothesis. *Economic Geography* 49: 68-73.

Haynes, K. E., and Ip, P. 1971. Population, economic development and structure of transportation in the province of Quebec, Canada. *Tijdschrift Voor Econ. en. Soc. Geographie* 62: 356-363.

Haynes, K. E., and Storbeck, J. 1978. The entropy paradox and the distribution of urban population. *Socio-Economic Planning Sciences* 12: 1-6.

Hodder, I. 1980. Spatial patterns of the past: problems and potential. In *Statistical Applications in the Spatial Sciences,* ed. N. Wrigley, pp. 189-202. New York: Methuen.

House, P. W., Patterson, P., and Williams, E. R. 1976. "Policymaking with divergent forecasts: A case study of data uses in federal decision making," in *Information Systems Studies.* Washington, DC: National Commission on Supplies and Shortages.

House, P. W., and Williams, E. R. 1978. Data inconsistencies and federal policymaking. *Policy Analysis* 8: 205-225.

Howrey, E. 1969. On the choice of forecasting models for air travel. *Journal of Regional Science* 9: 215-225.

Hoyt, H. 1958. *A re-examination of the shopping center market.* Urban Land Institute-Technical Bulletin, No. 33. Washington, DC: Urban Land Institute.

Huff, D. L. 1959. Geographical aspects of consumer behavior. *University of Washington Business Review* 18: 27-37.

———. 1960. A topographical model of consumer space preferences. *Papers and Proceedings, Regional Science Association* 6: 159-174.

———. 1961. Ecological characteristics of consumer behavior. *Papers and Proceedings, Regional Science Association* 7: 19-28.

———. 1962. *Determinatin of intra-urban retail trade areas.* Los Angeles: University of California, Graduate School of Business Administration.

———. 1963a. A probabilistic analysis of shopping center trade areas. *Land Economics* 39: 81-90.

———. 1963b. A probabilistic analysis of consumer behavior. *Papers and Proceedings of the Regional Science Association* 7: 81-90.

———. 1964. Defining and estimating a trade area. *Journal of Marketing* 28: 34-38.

———. 1966. A programmed solution for approximating an optimum retail location. *Land economics* 42: 293-303.

Hutchinson, B. G. 1974. *Principles of Urban Transport Systems Planning.*Washington, DC: Scripta.

Hutchinson, B. G., and Smith D. P. 1979. Empirical studies of the journey to work in urban Canada. *Canadian Journal of Civil Engineering* 6: 308-318.

Ikle, F. C. 1954. Sociological relationship of traffic to population and distance. *Traffic Quarterly* 8: 123-136.

Isserman, A. M., Plane, D. A., and McMillen, D. B. 1982. Internal migration in the U.S.: an evaluation of federal data. *Review of Public Data Use* 10: 285-311.

Jaynes, E. T. 1957. Information theory and statistical mechanics. *Physical Review* 105: 620-630.

Jochin, M. A. 1976. *Hunter-gatherers subsistence and settlement: a predictive model* New York: Academic Press.

Johnston, R. J. 1973. On frictions of distance and regression coefficients. *Area* 5: 187-191.

———. 1975. Map pattern and friction of distance parameters: a comment. *Regional Studies* 9: 281-283.

Kasakoff, A. B., and Adams, J. W. 1977. Spatial location and social organization: an analysis of tikopian patterns. *Man* 12: 48-64.

Keyfitz, N. and Philipov, D. 1981. The one-year/five-year migration problem. In *Advances in Multiregional Demography,* ed. Andrei Rogers, Laxenburg, Austria: International Institute for Applied Systems Analysis.

King, Leslie J. 1984. *Central Place Theory.* Beverly Hills, CA: Sage.

Knudsen, D. 1983. An analysis of temporary evolution in a spatial interaction system: a new look at U.S. commodity flows. Unpublished paper, Annual Meetings of the Association of American Geographers, Denver, Colorado.

Korcelli, P. 1975. Theory of urban spatial structure: review and synthesis. A cross-cultural perspective. *Geographia Polonica* 31: 99-131.

———. 1976. Urban spatial interaction models in a planned economy. *International Regional Science Review* 1: 74-87.

Lakshmanan, T. R., and Hansen, W. G. 1965. A retail market potential model. *Journal of the American Institute of Planners* 31: 134-143.

Lansing, J. B. et al. 1961. An analysis of interurban air travel. *Quarterly Journal of Economics* 75: 87-95.

Leinbach, T. R. 1973. Distance, information flows and modernization: some observations from West Malaysia. *The Professional Geographer* 25: 7-11.

Linneman, H. V. 1966. *An Econometric Study of International Trade Flows.* Amsterdam: North-Holland.

Long, W. H. 1968. City characteristics and the demand for interurban air trabel. *Land Economics* 44: 197-204.

———. 1969. Airline service and the demand for intercity air travel. *Journal of Transport Economics and Policy* 3: 287-299.

———. 1970. Air travel, spatial structure, and gravity models. *Annals of Regional Science* 4: 97-107.

Lösch, A. 1954. *The Economics of Location*(trans. from the revised ed. by W. H. Woglom and W. F. Stolper). New Haven, CT: Yale University Press.

Lowry, I. S. 1964: *A model of metropolis RM-4035-RC.* Santa Monica, CA: Rand Corporation.

———. 1967. *Seven models of urban development: a structural comparison.* Conference on Urban Development models, Dartmouth College, Hanover, NH. Highway Research Board, National Research Council.

Luce, R. D. 1959. *Individual choice behavior.* New York: John Wiley.

March, L. and Batty, M. 1975. Generalized measures of information, Bayes' likelihood ratio and Jayne's formalism. *Environment and Planning B* 2: 99-105.

Masser, I. 1972. *Analytical models for urban and regional planning.* Devon, England: David and Charles.

Meyer, J. R., and Straszheim, M. R. 1971. Forecasting demands for intercity freight transport. *Techniques of Transport Planning Vol. 1: Pricing and Project Evaluation* ed. J. R. Meyer. Washington, DC: Brookings Institute.

Morrill, R. L. 1959. *Service areas, highways and consumer movement: the example of physician care.* Department of Geography, Discussion Paper, No. 17. Seattle: University of Washington.

Morrill, R. L., and Pitts, F. R. 1967. Marriage, migration and the mean information field. *Annals of the Association of American Geographers* 57: 401-422.

Moses, L. N. 1955. The stability of interregional trading patterns and input-output analysis. *American Economic Review* 45: 803-826.

Nakanishi, M., and Cooper, L. G. 1974. Parameter Estimation for a multiplicative competitive interaction model-least squares approach. *Journal of Marketing Research* 11: 303-311.

Nelson, P. 1959. Migration, real income and information. *Journal of Regional Science* 1: 43-73.

Niedercorn, J. H., and Bechdolt, V. B., Jr. 1969. Economic derivation of the "gravity law" of spatial interaction. *Journal of Regional Science* 9: 273-282.

———. 1972. An economic derivation of the ''gravity law'' of spatial interaction: a further reply and a reformulation. *Journal of Regional Science* 12: 127-136.

Olson, A. L. 1972. A method for estimating regional redistributions of economic activity. *Papers of the Regional Science Association* 28: 181-187.

Openshaw, S. 1977. Timal zoning systems for spatial interaction. *Environment and Planning A* 9: 169-184.

Patton, D. J. 1956. *The demand for transportation: region and commodity studies for the U.S.* Research Paper, No. 45. Chicago: University of Chicago, Department of Geography.

Perle, E. 1964. *The Demand for Transportation*. Research Paper 95. Chicago: University of Chicago, Department of Geography.

Pfister, R. L. 1961. The terms of trade as a tool for regional analysis. *Journal of Regional Science* 3: 57-66.

Pitfield, D. E. 1978. Freight distribution model predictions compared: a test of hypotheses. *Environment and Planning A* 10: 813-836.

Plane, D. A. 1981. Estimation of place to place migration flows for net migration totals: a minimum information approach. *International Regional Science Review* 6: 33-52.

Polarczyk, K. 1976. The distribution of service centers within large urban areas. A market accessibility mode. *Geographia Polonica* 33: 143-155.

President's Commission on Federal Statistics. 1971. *Federal Statistic*. Washington, DC: Government Printing Office.

Putman, S. H. 1979. *Urban Residential Location Models.* Boston: Martinus Nijhoff.

Quandt, R. E. 1965. Some perspectives of gravity models. *Studies in Travel Demand.* Washington, DC: Department of Commerce.

Ravenstein, E. G. 1885 & 1889. The laws of migration. *Journal of the Royal Statistical Society* 48: 167-235; 52: 241-305.

Reilly, W. J. 1929. *Methods for the study of retail relationships.* University of Texas Bulletin, No. 2944. Austin: University of Texas.

Richmond, S. B. 1955. Forecasting air passenger traffic by multiple regression analysis. *Journal of Air Law and Commerce* 22: 434-443.

———. 1957. Interspatial relationships affecting air travel. *Land Economics* 33: 65-73.

Riddell, J. B. 1968. *Structure, diffusion and response: the spatial dynamics of modernization in Sierre Leone.* Ph.D. dissertation, Department of Geography, the Pennsylvania State University.

Rogerson, P. A., and Plane, D. A. Modeling temporal change in flow matrices. *Papers of the Regional Science Association*, in press.

Rosander, C. 1965. Obtaining acceptable quality data from carload waybill and often sampler. Highway Research Record 82: 114-120.

Shershevsky, I. Z. and Shnurov, M. C. 1965. K rashchotu optymalnego rasselenya s primeneniyu elektronno-cifrowykh vychislitelnykh mashin. *Gorodoskoye Khazaystro Moskvy* 2.

Senior, M. L. 1979. From gravity modeling to entropy maximizing: a pedagogic guide. *Progress in Human Geography* 3: 175-210.

Smith, T. R., and Clayton, C. 1978. Transitivity, spatial separable utility functions and United States migration streams, 1935-1970. *Environment and Planning A* 10: 399-414.

Smith, T. R., and Slater, P. B. 1981. A family of spatial interaction models incorporating information flows and choice set constraints applied to U.S. interstate labor flows. *International Regional Science Review* 6: 15-31.

Sokolov, V. 1975. Models aiding national settlement policies in the USSR: a survey. *Environment and Planning A* 7: 757-780.

Sosslau, A. B. et al. 1978. *Quick-response urban travel estimation techniques and transferable parameters: user's guide.* Washington, DC: Transportation Research Board, National Research Council, U.S. Department of Transportation.

Southworth, F. 1982. *A scenario-based approach to assessment of requirement for freight transportation,* Map 2000 Project. Metropolitan Housing and Planning Council Report, No. 16. Chicago, Illinois.

Southworth, F., Lee, Y. J. , Griffin, C. S., and Zavattero, D. 1983a. *Motor freight planning for the Chicago region.* Illinois Department for Energy and Natural Resources Doc., No. 83/06. Springfield, IL: Author.

———. 1983b. *Strategic motor freight planning for the year 2000, Chicago.* Transportation Planning Group Publication, No. 22. Urbana: Department of Civil Engineering University of Illinois.

Spiegelglas, S. 1960. Some aspects of state-to-state commodity flows in the United States. *Journal of Regional Science* 2: 71-80.

———. 1940. Intervening opportunities: a theory relating mobility and distance. *American Sociological Review* 5: 845-850.

Stewart, J. Q. 1940. The gravity of the Princeton family. *Princeton Alumni Weekly* 40: 14-29.

Stillwell, J. C. H. 1978. Interzoned migration: some historical tests of spatial-interaction models. Environment and Planning A 10: 1187-1200.

Stouffer, A. 1960. Intervening opportunities and competing migrants. *Journal of Regional Science* 1: 1-20.

Stutz, F. P. 1973. Distance and network effects on urban social travel fields. *Economic Geography* 49. 134-144.

———. 1976. *Social aspects of interaction and transportation.* A.A.G. Resource Paper No. 76-2. Washington, DC: Association of American Geographers.

Taaffe, E. J. 1956. Air transportation and the U.S. urban distribution. *Geographical Review* 46: 219-238.

———. 1959. Trends in airline passenger traffic. *Annals of the Association of American Geographers* 49: 560-575.

Tobler, W., and Wineberg, S. 1971. A Cappadocean speculation. *Nature* 321 (5297): 39-42.

Tobler, W. 1983. An alternative formulation for spatial-interaction modeling. *Environment and Planning A* 15: 693-703.

Tornquist, G. 1968. *Flows of information and the location of economics activity.* Lund Series in Geography Series B, No. 30. Gleerup: Lund.

Trudgill, H. 1975. Linguistic geography and geographical linguistics. *Progress in Geography* 7: 227-252.

Ullman, E. L. 1957. *American Commodity Flow.* Seattle: University of Washington Press.

Urban Mass Transit Administration. 1972-78. *UMTA transportation planning systems (UTPS) reference manuals.* Washington, DC: U.S. Department of Transportation.

U.S. Bureau of Economic Analysis. 1976. *Regional work force characteristics and migration data: a handbook on the social security continuous work history sample and its application.* Washington, DC: U.S. Government Printing Office.

U.S. Department of Commerce. 1965. *Fitting the gravity model for any urban size.* Washington, DC: Author.

U.S. Department of Transportation. 1973. *Calibrating and testing a gravity model for any urban size.* Washington, DC: Federal Highway Administration.

U.S. General Accounting Office. 1974. *U.S. actions needed to cope with commodity shortages.* Washington, DC: Comptroller General.

Webber, M. J. 1977. Pedagogy again: what is entropy. *Annals of the Association of American Geographers* 67: 254-266.

———. 1980. *Information Theory And Urban Spatial Structure.* London: Croom Helm.

Wheeler, J. O., and Stutz, F. 1971. Spatial dimensions of urban social travel. *Annals of the Association of American Geographers* 61: 371-386.

Wheeler, J. O. 1972. Trip purposes and urban activity linkage. *Annals of the Association of American Geographers* 62: 641-654.

Wilson, A. G. 1967. A statistical theory of spatial distribution models. *Transportation Research* 1: 253-269.

———. 1971. A family of spatial interaction models, and associated developments. *Environment and Planning A* 3: 1-32.

Young, E. C. 1924. *The movement of farm population.* Agricultural Experimental Station Bulletin Ithaca. NY: Cornell University.

Young, W. J. 1975. Distance decay values and shopping center size *Professional Geographer* 27: 304-309.

Zagozdzon, A. 1973. Problems of development of a settlement network in a region under industrialization. *Geographia Polonica* 27: 159-174.

———. 1976. Regional and subregional centers of Poland: a general characterization. *Geographia Polonica* 33: 175-189.

Zipf, G. K. 1949. *Human Behavior And The Principle Of Least Effort.* Reading, MA: Addison-Wesley.

Zipser, T. 1973. A simulation model of urban growth based on the model of opportunity selection, *Geographia Polonica* 27: 119-132.

ABOUT THE AUTHORS

KINGSLEY E. HAYNES is Professor of Geography and Professor of Public and Environmental Affairs at the School of Public and Environmental Affairs, Indiana University, Bloomington. He is also Director of the Center for Urban and Regional Analysis and Chairman of the Urban, Regional Analysis and Planning Faculty. He received his Ph.D. (1971) in geography and environmental engineering from Johns Hopkins University, his M.A. (1965) in geography from Rutgers University, and his B.A. (with honors) in geography, history and political science from Western Michigan University in 1964. Professor Haynes has published over one hundred articles on many different subjects—regional settlement patterns, information theoretic methods, hierarchical goal programming, multiobjective location analysis and energy facility siting. He is also recognized for his applications of spatial modeling to such real world problems as management strategies for operating the Aswan High Dam and development of the Upper Nile Basin, resource use strategies for Malaysian development alternatives, airport site impacts for the civil aviation authority in Brazil, and siting of the Montreal International Airport for the Canadian Department of Transportation.

A. STEWART FOTHERINGHAM is Associate Professor of Geography at the University of Florida. Between 1980 and 1984 he was an Assistant Professor of Geography at Indiana University, Bloomington. He received his Ph.D. (1980) and M.A. (1978) in geography from McMaster University, Canada. He received his B.Sc. (with honors) from the University of Aberdeen, Scotland, in 1976. Professor Fotheringham was a joint recipient of the Warren G. Nystrom Award at the Seventy-Seventh Annual Meeting of the Association of American Geographers. This award recognizes an outstanding geography dissertation published the previous year; his dissertation focused on gravity and spatial interaction models. Professor Fotheringham has published in such journals as *Annals of the Association of American Geographers, Environment and Planning, Geographical Analysis,* and *Economic Geography.*